Diet, Nutrition, and Cancer: A Critical Evaluation

Volume I
Macronutrients and Cancer

Editors

Bandaru S. Reddy, Ph.D.
Member and Associate Chief, Division of
Nutrition and Endocrinology
Head, Nutritional Biochemistry Section
Naylor Dana Institute for Disease Prevention
American Health Foundation
Valhalla, New York

Leonard A. Cohen, Ph. D.
Head, Nutritional Endocrinology Section
Division of Nutrition and Endocrinology
Naylor Dana Institute for Disease Prevention
American Health Foundation
Valhalla, New York

CRC Press
Taylor & Francis Group
Boca Raton London New York

CRC Press is an imprint of the
Taylor & Francis Group, an **informa** business

Library of Congress Cataloging in Publication Data
Main entry under title:

Diet, nutrition, and cancer.

  Includes bibliographies and indexes.
  Contents: v. 1. Macronutrients and cancer --
v. 2. Micronutrients, nonnutritive dietary factors,
and cancer.
  Cancer--Nutritional aspects. I. Reddy,
Bandaru S. II. Cohen, Leonard A. [DNLM: 1. Diet--
adverse effects. 2. Neoplasms--etiology. 3. Neoplasms
--prevention & control. 4. Nutrition. QZ 202 D5653]
RC268.45.D54 1986   616.99'4   85-15172
ISBN 0-8493-6332-2 (v. 1)
ISBN 0-8493-6333-0 (v. 2)

A Library of Congress record exists under LC control number: 85015172

ISBN 13: 978-1-315-89230-6 (hbk)
ISBN 13: 978-1-351-07140-6 (ebk)

Visit the Taylor & Francis Web site at http://www.taylorandfrancis.com and the
CRC Press Web site at http://www.crcpress.com

PREFACE

Although the concept that diet and nutrition might influence cancer is not new, until recently this relationship has received surprisingly little detailed attention. During the 1930s, a number of laboratories, including that of Tannenbaum, were interested in the possible influence exerted by nutritional factors on susceptibility to cancer, but the question soon lost the interest of both the scientific and lay community.

During the past 2 decades, renewed interest in nutritional carcinogenesis has developed. Epidemiologic studies have investigated the incidence pattern between and among population groups, differences in the rates of the disease between the sexes, changes in disease rates over time, demographic and socioeconomic distribution of diseases, effects of migration, and the dietary habits of different population groups, and have led to the conclusion that nutritional factors play a significant role in the etiology of certain types of cancer. However, it must be recognized that the correlation between nutritional factors and certain forms of cancer does not prove causation. Many factors may be necessary for cancer causation, but the modification of only one of the contributing factors, such as diet, may be sufficient to retard the chain of causative events.

Studies in experimental animal models also point to dietary factors as important modulators of certain types of cancer. These studies have generally shown that increased macronutrient intake, especially fat, and certain micronutrient deficiencies lead to increased in tumor incidence in several organ sites, whereas diet restriction and dietary excess of certain micronutrients lead to a lower tumor incidence.

These 2 volumes bring together a wide variety of studies concerning the role nutrition plays in the etiology of various types of cancer, namely, cancer of the esophagus, upper alimentary tract, pancreas, liver, colon, breast, and prostate. The purpose of each chapter is to provide a critical interpretive review of the area, to identify gaps and inconsistencies in present knowledge, and to suggest new areas for future reasearch. Scientifically valid data supporting an association between nutrition and cancer comes from three sources: epidemiology, clinical studies, and experimental studies in laboratory animal models. Throughout the volumes, attention is given to the potential and limitations of each discipline; and the need for closer cooperation between epidemiologists, clinicians, and experimentalists is emphasized. Specific areas of concern include extrapolation of data from animal models to humans, methods of diet evaluation, the formation and occurrence of mutagens in cooked food, and the role of naturally-occurring inhibitors of carcinogenesis.

We have tried to present in 19 chapters (9 chapters in Volume I, 10 chapters in Volume II) a comprehensive view of nutrition's role in cancer. The broad coverage of diet, nutrition, and cancer provided by these chapters is intended to serve both as an introduction to readers unfamiliar with the field, as well as a source of new information for researchers. It is indeed hoped that these 2 volumes will promote a better understanding of the role of nutritional factors in the induction and inhibition of cancer, and that this understanding will lead to a reduction in cancer rates in the current generation and the prevention of cancer in future generations.

Obviously, the compilation of these volumes could not have come about without the cooperation of the various authors. We most sincerely thank each of the authors for their contribution and continued assistance in submitting and editing the manuscript.

THE EDITORS

Bandaru S. Reddy, D.V.M., Ph.D., is presently a Member and Associate Chief of the Division of Nutrition and Endocrinology, Naylor Dana Institute for Disease Prevention, Valhalla, New York, and Research Professor of Microbiology at New York Medical College, Valhalla, New York. He received a degree in Veterinary Medicine in 1955 from the University of Madras, India, an M.S. in 1960 from the University of New Hampshire, and Ph.D. in 1963 from Michigan State University. He then spent 8 years as a faculty member at the Lobund Laboratory of the University of Notre Dame, Notre Dame, Indiana.

His current research interest is diet, nutrition and cancer, with particular emphasis on large bowel cancer. This is an important area of investigative research that may contribute to primary and secondary prevention of cancer. He has published about 160 papers on this and related subjects.

Dr. Reddy is a member of the American Institute of Nutrition, American Association of Pathologists, American Association of Cancer Research, Society for Experimental Biology and Medicine, Association of Gnotobiotics, and the Society of Toxicology. During 1979, he served as the President of the Association of Gnotobiotics.

Leonard A. Cohen is presently Head, Section of Nutritional Endocrinology, Naylor Dana Institute for Disease Prevention, Valhalla, New York. He received a Ph.D. from the City University of New York in 1972 in cell biology. Dr. Cohen joined the NDI soon after receiving his doctorate and has dedicated his efforts since then to the elucidation of the mechanism of mammary tumor promotion by dietary fat.

Dr. Cohen is a member of the American Association for Cancer Research, the International Association for Breast Cancer Research, The International Association for Vitamins and Nutritional Oncology, The Tissue Culture Association, and the American Association for the Advancement of Science.

CONTRIBUTORS

Maarten C. Bosland
Research Assistant Professor
Institute of Environmental Medicine
New York University Medical Center
New York, New York

Andrea P. Boyar
Head, Section of Clinical Nutrition
Division of Nutrition and Endocrinology
Mahoney Institute for Health
  Maintenance
American Health Foundation
New York, New York

Leonard A. Cohen
Head, Section of Nutritional
  Endocrinology
Division of Nutrition and Endocrinology
Naylor Dana Institute for Disease
  Prevention
American Health Foundation
Valhalla, New York

Pelayo Correa
Professor of Pathology
Louisiana State University
New Orleans, Louisiana

David F. Horrobin
Director
Efamol Research Institute
Nova Scotia, Canada

Ole Møller Jensen
Director
Danish Cancer Registry
Institute of Cancer Epidemiology
Under the Danish Cancer Society
Copenhagen, Denmark

Daniel S. Longnecker
Professor of Pathology
Department of Pathology
Dartmouth Medical School
Hanover, New Hampshire

A. B. Miller
Director
NCIC Epidemiology Unit
University of Toronto
Toronto, Ontario
Canada

Reginald G. H. Morgan
Associate Professor of Pathology
Department of Physiology
The University of Western Australia
Nedlands, Western Australia

Bandaru S. Reddy
Associate Chief, Division of Nutrition
  and Endocrinology
Naylor Dana Institute for Disease
  Prevention
American Health Foundation
Valhalla, New York

David P. Rose
Chief
Division of Nutrition and Endocrinology
Naylor Dana Institute for Disease
  Prevention and Mahoney Institute for
  Health Maintenance
American Health Foundation
New York, New York

TABLE OF CONTENTS

TABLE OF CONTENTS

Chapter 1

DIET AND GASTRIC CANCER

Pelayo Correa

TABLE OF CONTENTS

I. INTRODUCTION

There is a general consensus implicating diet as the main factor in human gastric cancer etiology. This concept has developed over the years and is based mostly on circumstantial evidence. There is, however, no scientific proof for it and no general agreement on the specific components of the diet supposedly responsible for gastric cancer.

Extensive reviews of the pertinent literature are available.[1-4] Rather than repeating or updating such reviews, this chapter attempts to screen the relevant literature and select out those items which appear to have resisted the test of time and survived as viable candidates in the chain of events which may eventually lead to gastric cancer. Finally, their specific role in a proposed etiologic model will be considered.

II. WHY DIET?

Ingested materials are first detained and acted upon by digestive enzymes in the stomach. It is, therefore, not surprising that diet is considered a prime candidate for gastric carcinogenesis. Several mechanisms have been mentioned to explain the role of diet in gastric carcinogenesis: (1) presence of carcinogens in food; (2) introduction of carcinogens during food preparation; (3) absence of protective factors; (4) synthesis of carcinogens by interaction of food items; (5) irritants in food resulting in cancer promotion.[2,5]

Some support for the speculation has been provided by the geographic distribution of the disease. Figure 1 is based on available data on gastric cancer incidence. The intercounty contrast is very prominent but geography by itself cannot explain why some subpopulations display risks several times greater than others inhabiting the same land. Chinese have rates several times greater than Malays in Singapore. Similar contrasts are observed between Maoris and whites in New Zealand; Japanese and whites in Hawaii; Indians and whites in New Mexico; blacks and whites in California; and Jews and Arabs in Israel.

In the contrasting situations mentioned above, racial and cultural differences exist between subpopulations inhabiting the same land. The role of race has been studied in migrant populations. Immigrants from high-risk areas to generally low-risk environment of the U.S. display slightly lower risks of gastric cancer in the first generation, but the second generation displays a dramatic drop in their risk.[6] In some racial groups, like the Japanese, the risk reduction cannot be explained by interracial marriages. Similar experiences have been reported for immigrants to Austrialia, Canada, and Brazil.

Descriptive epidemiology studies, therefore, clearly indicate that race and geography are not the main determinants of gastric cancer frequency. Culture is strongly implicated. One of the main cultural differences between subpopulations at high and low risk is the diet. No contrast in gastric cancer risk is on record among populations with similar diet.

III. THE HIGH-RISK DIET

Great diversity exists between the diets of populations at high gastric cancer risk. This probably explains why there was so much discrepancy in the results of the earlier dietary studies in several countries: rice was suspected in Japan, fried foods in Wales, potatoes in Slovenia, grain products in Finland, spices in Java, and smoked fish in Iceland.[2] More recent studies have emphasized salty foods.[7-9] The most remarkable trend, however, has been the realization that some old and some new studies are more prominently showing negative associations than positive associations.[7-8, 10-13] The analytical epidemiologic data, therefore, strongly point to the presence of protective factors in the diet.

Looking for common factors in the diet of high-risk populations and taking note of descriptive, correlational, and analytical studies, the ''gastric cancer diet'' has been characterized as follows:[14]

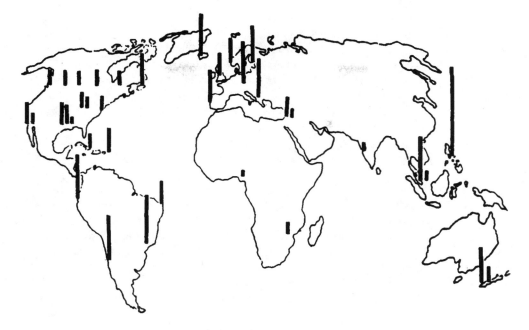

FIGURE 1. Diagram representing age-adjusted incidence rates of stomach cancer in different populations. (From *Cancer Incidence in Five Continents*, Vol. 4, International Agency for Research on Cancer, Lyon, 1976.)

1. Low in animal fat and animal proteins
2. High in complex carbohydrates
3. A substantial proportion of the protein is obtained from vegetable sources, mostly grains
4. Low in salads and fresh, green, leafy vegetables
5. Low in fresh fruits, especially citrus
6. High in salt

An additional, somewhat controversial, item which should be considered is the presence of nitrates in the diet. The evidence implicating the above items is reviewed in the following paragraphs.

A. Animal Fats and Proteins

Most populations at high gastric cancer risk have diets low in animal fats and proteins. The protective effect of milk has been found prominently in Japan.[7] Fat is needed for the proper utilization of lipid-soluble vitamins such as A and E, which may play a protective role. There are, however, many populations whose intake of animal fat and proteins is very low and still display very low risk of gastric cancer. Such is the case of most African aboriginals and most inhabitants of the tropical and subtropical lowlands, including most of Brazil and the Caribbean basin.

The evidence available cannot totally deny a possible role of low-fat diet in some situations but does indicate that other factors are needed to induce gastric cancer in humans.

B. Abundance of Complex Carbohydrates

The situation concerning this item is similar to the low-fat diet: it applies to most high-risk populations but it is equally prevalent in large populations displaying low risk. There appears to be a distinction, however, in the type of vegetables frequently consumed. In most situations high-risk populations base their diet on grains and roots which are subjected to involved cooking techniques. These populations are not inclined to eat fresh, uncooked vegetables. Not enough data exist to determine cultural differences between high- and low-risk populations with a predominantly vegetable diet. One recent study in Colombian rural villages found that a vegetable-based diet was predominant in all villages, but only the low-risk villagers were addict to fresh items (fruits and salads).[14] The abundance of complex carbohydrates, therefore, is not a strong determinant factor in gastric carcinogenesis. The protective role of this type of diet in colon carcinogenesis is discussed elsewhere, but it should be mentioned here that large populations with high intake of complex carbohydrates have very low rates of both gastric and colon cancer: Brazil, Africa, and the Caribbean.

C. Salads and Fruits

The negative association between these items and gastric cancer is one of the most consistent findings: old and new studies in populations of different race and culture coincide on this point. Stocks reported it in England in 1933;[15] Dunham in 1946, Haenszel in 1958, Higginson in 1959, and Graham in 1972 in the U.S.;[11,16-18] Meinsma in Holland 1964;[13] Hirayama in Japan in 1963;[7,19] Paymaster in India in 1968;[20] Bjelke in Norway in 1970;[21] and Haenszel in Hawaii in 1972.[8] The specific kind of fresh fruit or vegetable associated with low risk varies according to cultural dietary patterns, but it appears obvious that some common factor or factors exist in many fresh fruits and vegetables that may play a protective role in gastric carcinogenesis. The search for the common factor has led investigators to estimate the intake of micronutrients based on their concentration on specific food items, the usual size of a typical serving of each food item, and the frequency with which the item is reportedly eaten by cases and controls. The result of these calculations is usually expressed as an "index".[22] Such indexes may be adequate for hypothesis building but not for hypothesis testing because they are fraught with potential errors in the basic assumptions. Marked differences in the concentration of a micronutrient in a specific food may result from soil composition and fertilization patterns: the concentration of vitamin C in grains is considerably greater when in soils fertilized with ammonium molybdate than in those deficient in that element.[23] Most dietary surveys have to limit the number of dietary items for logistic reasons; it is entirely possible that some items rich in a specific micronutrient are left out of the questionnaire. The dietary questionnaires have to cover periods prior to clinical disease, making quantitation only reliable for rather gross categories. Indexes of micronutrients based on case-control studies have been calculated for ascorbic acid[21] and vitamin A.[24] In Japan, a large cohort study has revealed a negative association between green-yellow vegetable intake and gastric cancer risk; such vegetables account for approximately 50% of vitamin A intake and 25% of vitamin C intake.[25] Similar calculations tend to implicate vitamin E in gastric cancer precursors.[14] All indexes, however, show an association with gastric cancer risk which is somewhat weaker than the direct association with the intake of fresh fruits and vegetables, from which they derive. This may indicate that more than one micronutrient is involved in the process. Indexes of micronutrients may, on the other hand, be indicators of other, nonnutrient, protective factors in food. The list of such factors is large and so far insufficiently explored.[26-27]

D. Salt

It has long been suggested that high salt concentration in the diet increases gastric cancer risk. Sato reached that conclusion in 1959 after correlating gastric cancer death rates in Japanese prefectures with local customs of salting foods, especially vegetables.[28] Similar results have been reported in more recent surveys.[29] The hypothesis was tested and found correct in case-control studies in Japan,[7,30] in France,[31] and in Hawaii.[8] Extensive international correlation studies between death rates from cerebrovascular accidents and gastric cancer has led Joosens to conclude that both are related to excessive salt intake.[32] Similar conclusions have been reached in Colombia for gastric cancer precursors.[14] One recent Japanese report failed to detect an association between salt intake and prefecture-specific gastric cancer death rates, probably reflecting recent dietary changes in Japan which may affect future (but not present) death rates. Epidemiologically, the case for a role of excessive salt intake in gastric carcinogenesis appears very strong. This conclusion is reinforced by experimental studies showing that salted food causes severe gastritis (a strongly suspected cancer precursor) in man[33] and in experimental animals.[34] The hypothesis has also been tested in experimental carcinogenesis studies. It has been found that salt increases the yield and size of tumors of the glandular stomach of Wistar rats when administered before or during MNNG administration. Salt alone is not a complete carcinogen or an initiator. When given after the carcinogen MNNG it is ineffective and therefore can not be called a promoter.[35,36] It does appear that salt is an etiologic factor of gastric cancer that does not conform to the stereotypes of the initiator-promoter classification. Its role may be in inducing chronic gastritis (which increases cell turnover rates) and facilitating the contact of the carcinogens with their target cells, probably by disruption of the mucus barrier.

E. Nitrates and Nitrites

The extensive literature linking nitrate and nitrite to gastric cancer has been recently reviewed.[37] A positive intercountry correlation was found between gastric cancer death rates and estimates of nitrate consumption in the diet. Correlations between the use of nitrate-rich well water and gastric cancer risk has been found in Colombia.[38] A correlation between nitrates in water supplies and gastric cancer rates was reported from England[39] but later interpreted as reflecting socioeconomic gradients and occupational patterns in the country.[40] A model for gastric carcinogenesis has been proposed, which postulates intragastric bacterial reduction of nitrate to nitrite and subsequent synthesis of carcinogenic nitroso compounds.[5] Nitrate itself is not considered a precursor of carcinogens because it will not react with amines or other nitrogen-containing molecules to form N-nitroso compounds. Nitrite, on the other hand, is a strong carcinogen precursor because it avidly nitrosates amines, amides, ureas, and similar compounds.

What matters, therefore, is the exposure to nitrites in the gastric cavity. Most of the nitrite exposure in humans results from reduction of dietary nitrate, but nitrate is present in all human diets in quantities more than adequate to provide high nitrite levels if subjected to bacterial reduction. It is pointless, therefore, to attempt to reduce nitrate levels in the diet.

Dietary nitrate as well as nitrate excreted by the salivary glands is in part reduced to nitrite in the buccal cavity. This has led to the study of salivary nitrate and nitrite as possible indicators of gastric cancer risk with negative results.[41] Salivary nitrate and nitrite levels in some populations may reflect the intake of fresh vegetables (rich in nitrate) and therefore result in a negative association of gastric cancer risk. Salivary nitrite is not an indicator of nitrite levels in the gastric cavity. The latter is most probably determined predominantly by the presence of reductase-containing bacteria in the gastric cavity, associated with chronic atrophic gastritis and bacterial proliferation due to partial loss of gastric acidity.[5] It has recently been shown that a given dose of nitrate administered to patients with chronic atrophic gastritis results in concentrations of nitrite in the gastric juice that are approximately ten times highter than those of normal subjects receiving the same dose of nitrate.[42] Gastric microenvironment (not buccal microenvironment) is therefore the pertinent parameter.

The role of nitrates and nitrites in gastric carcinogenesis cannot be considered proven. In part, it hinges on the etiologic hypothesis postulating intragastric nitrosation of nitrogen-containing compounds. The hypothesis is not proven but has received considerable support. It should be, however, understood that dietary nitrate is not the key etiologic factor. Reduction of dietary nitrate to nitrite in the gastric cavity is the key issue.

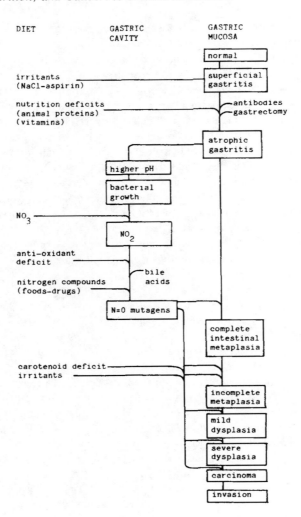

FIGURE 2.   Diagramatic representation of the etiologic hypothesis of gastric cancer.

The intragastric nitrosation hypothesis is supported abundantly by laboratory evidence. A direct mutagen has been found in the gastric juice of patients with high nitrite levels.[43] High levels of nitrites and nitrosamines have been reported in the gastric juice of patients with atrophic gastritis.[44] Gastric tumors have been induced in rodents by simultaneous administration of nitrite and amines.[45] Fish commonly used in Japanese diet can be nitrosated to produce carcinogenic substances which induce experimental tumors in rodents.[46]

IV. ETIOLOGIC HYPOTHESIS

Information on gastric carcinogenesis, from several scientific disciplines, has been utilized to postulate an etiologic model which has been recently updated.[5,47] The hypothesis centers around the gastric microenvironment playing a key role in the pathogenesis of gastric cancer. The microenvironment is determined by three basic elements: what is ingested (diet), what is secreted or produced locally in the gastric cavity, and what changes take place in the target organ: the gastric mucosa. The hypothesis is outlined in Figure 2.

In this hypothesis diet plays an overriding role since it can have a decisive influence on the gastric secretions and the gastric mucosa itself. The first element of the diet to be considered is the presence of irritants which can damage the gastric mucosa directly. The most important of those irritants is probably salt, as discussed above. Other irritants of importance are alcohol and aspirin. Irritants are probably responsible for the chronic gastritis, a precursor of gastric cancer.

The next step in the process is that of atrophy, one of the least understood steps in the chain of precursor lesions. Whether diet plays a role in gastric atrophy is unknown. In some high-risk populations protein supply is indadequate, which could interfere with proper regeneration of epithelial cells lost because of the chronic gastritis process. Gastrectomy and pernicious anemia are causes of atrophy not related to diet.

Once atrophic gastritis sets in, drastic changes take place in the microenvironment: the pH is progressively elevated and bacterial proliferation occurs in the lumen and in the secretions immediately covering the epithelial cells. The bacterial growth in the lumen is well documented and includes several species with nitrate reductases.[48] Nitrate present in the food is then reduced by bacteria and nitrite appears in the gastric secretions. To what extent bacteria are also present in the mucosal tissue itself is not well documented. Nitrite is a very active molecule which can react with a large number of nitrogen-containing species to produce *N*-nitroso compounds, many of which are mutagenic and carcinogenic. This is a central theme in the hypothesis and it has received support from several recent findings, summarized below.

In spite of the fact that nitrite disappears from the stomach soon after its formation, relatively high levels of nitrite are found in the gastric juice of persons at high gastric cancer risk,[49] incuding patients with pernicious anemia[50] and post-Bilroth II gastrectomy.[51] A direct correlation has been found between increasing pH and nitrite level of fasting gastric juice. High levels of total nitrosamines and of volatile nitrosamines have been reported in the gastric juice of high-risk populations.[52] Some questions remain unanswered, however, concerning the meaning and relevance of this work since volatile nitrosamines are not expected to be carcinogenic unless metabolized by the liver. Mutagens have been detected in the gastric juice of patients with atrophic gastritis.[43] Several food items frequently eaten by high-risk populations yield potent mutagens after nitrosation. Japanese fish after nitrosation yield a mutagen which has induced gastric carcinomas in experimental animals when administered via gastric tube.[46] In Colombia, fava beans are consumed abundantly by some high-risk populations. When nitrosated, fava beans yield a potent mutagen.[53]

Other candidates for nitrosation are bile acids. *N*-nitroso bile acids are mutagens and can be formed under simulated gastric conditions.[54] These compounds could explain the so-called stump carcinoma of post-Bilroth II gastrectomy patients.[55] Experimentally, gastric cancer can be induced by Bilroth II gastrectomy or gastroentrostomy without the administration of a carcinogen.[56]

Many drugs can be nitrosated and some may yield mutagens after nitrosation. Frequently used blockers of gastric secretion such as cimetidine and ranitidine yield mutagens after nitrosation but there is no proof that the products are carcinogenic.[57] Some pre-anesthesia sedatives have been reported to undergo nitrosation and yield mutagens within the gastric mucosa itself, in areas of intestinal metaplasia.[58] This may indicate that nitrate is absorbed into the mucosa from the gastric lumen or that bacterial colonization of the mucosa itself has taken place. That nitrosation does take place in vivo has been demonstrated by the so-called Bartsch test based on the detection of nitroso-proline in the urine after ingestion of nitrite and proline.[59]

The diet of populations at high risk is generally deficient in fresh fruits and vegetables. The meaning of this well-established cross-cultural finding is not well understood, especially since those food items contain hundreds of pro- and anticarcinogenic compounds whose effects in human populations is almost completely unknown.[26] The interpretations of the above findings have concentrated on two hypotheses: deficiency of antioxidants and deficiency of carotenoids. Antioxidant deficiency is suspected to play a role because of their well-known capacity to block nitrosation. The best-known compounds of this kind are ascorbic acid which acts in hydrosoluble mixtures and alpha-tocopherol which acts in liposoluble mixtures.[60] Many other compounds efficiently block nitrosation, such as caffeic acid found in coffee and propylgallic used for food preservation.[61] Ascorbic acid may also have an anticancer role even after the invasive state has been reached: it promotes fibroplasia and apparently creates obstacles for the spread of tumors.[62]

Carotenoid deficiency has been implicated in many epidemiologic studies, not only in relation to gastric cancer, but to several other epithelial tumors.[63] Their mechanism of action is not understood. It is suspected to be related to their conversion to retinol, a promoter of differentiation, and as such may interfere with the gradual loss of differentiation postulated in the model, especially the stages after complete metaplasia. Extensive evidence supports the preventive role of retinoids,[64] but little experimental work on carotenoids is available.

The model suggests that although the basic mechanisms may be similar, they may be operating with different ingredients in populations at high risk to gastric cancer. Although diet may be etiologically implicated, it may reach the same point with different components: for instance, it may be that a carcinogen from nitrosated fish plays a role in Japan while a carcinogen from nitrosated fava beans plays a role in Colombia. Prevention, therefore, may be tailored to the needs of each population.

REFERENCES

1. **Bjelke, E.,** Epidemiologic studies of cancer of the stomach, colon and rectum, *Scand. J. Gastroenterol.,* 9 (Suppl. 31), 164, 1974.
2. **Haenszel, W. and Correa, P.,** Developments in the epidemiology of stomach cancer over the past decade, *Cancer Res.,* 35, 3452, 1975.
3. **Barrett, M. K.,** Avenues of approach to gastric cancer problem, *J. Natl. Cancer Inst.,* 7, 127, 1946.
4. **Doll, R.,** Environmental factors in the aetiology of cancer of the stomach, *Gastroenterologia,* 86, 320, 1956.
5. **Correa, P., Haenszel, W., Cuello, C., Archer, M., and Tannenbaum, S.,** A model for gastric cancer epidemiology, *Lancet,* 2, 58, 1975.
6. **Haenszel, W.,** Cancer mortality among the foreign-born in the United States, *J. Natl. Cancer Inst.,* 26, 37, 1961.
7. **Hirayama, T.,** The epidemiology of cancer of the stomach in Japan, with special reference to the role of diet, *Gann Monogr.,* 3, 15, 1968.
8. **Haenszel, W., Kurihara, M., Segi, M., and Lee, E. K. C.,** Stomach cancer among Japanese in Hawaii, *J. Natl. Cancer Inst.,* 49, 969, 1972.
9. **Joossens, J. V. and Geboers, J.,** Nutrition and gastric cancer, *Nutr. Cancer,* 2, 250, 1981.
10. **Stocks, P.,** Cancer incidence in North Wales and Liverpool Region in Relation to Habits and Environment, Supplement to Part II of British Empire Cancer Campaign 35th Annual Report Covering the year 1957.
11. **Haenszel, W.,** Variation in the incidence of and mortality from stomach cancer, with particular reference to the United States, *J. Natl. Cancer Inst.,* 21, 213, 1958.
12. **Graham, S., Lilienfeld, A. M., and Tidings, J. E.,** Dietary and purgation factors in the epidemiology of gastric cancer, *Cancer,* 20, 2224, 1967.
13. **Meinsma, L.,** Voeding en kanker. Voeding 25:357-365, 1964.
14. **Correa, P., Cuello, C., Fajardo, L. F., Haenszel, W., Bolaños, O., and deRamirez, B.,** Diet and gastric cancer: nutrition survey of a high-risk area, *J. Natl. Cancer Inst.,* 70, 673, 1983.
15. **Stocks, P. and Karn, M. K.,** A co-operative study of the habits, home life, dietary and family histories of 450 cancer patients and of an equal number of control patients, *Ann. Eugen. (London),* 5, 137, 1933.
16. **Dunham, L. J. and Brunschwig, A.,** A review of dietary and related habits in patients with malignant gastric neoplasms, *Gastroenterology,* 6, 286, 1946.

17. **Higginson, J.**, Etiological factors in gastro-intestinal cancer in man, *J. Natl. Cancer Inst.*, 37, 527, 1966.
18. **Graham, S., Schotz, W., and Martino, P.**, Alimentary factors in the epidemiology of gastric cancer, *Cancer*, 30, 927, 1972.
19. **Hirayama, T.**, A study of the epidemiology of stomach cancer, with special reference to the effect of diet factor, *Bull. Inst. Publ. Health*, 12, 85, 1963.
20. **Paymaster, J. C., Sanghvi, L. D., and Gangadharan, P.**, Cancer in the gastrointestinal tract in Western India. Epidemiologic study, *Cancer*, 21, 279, 1968.
21. **Bjelke, E.**, Case-control study of cancer of the stomach, colon and rectum, in *Oncology 1970: Being the Proceedings of the Tenth International Cancer Congress*, Vol. 5, Clark, R. L., Cumley, R. W., McCay, J. E., and Copeland, M. M., Eds., Yearbook Medical Publishers, Chicago, Ill., 1971, 320.
22. **Marshall, J., Graham, S., Mettlin, C., Shedd, D., and Swanson, M.**, Diet in the epidemiology of oral cancer, *Nutr. Cancer*, 3, 145, 1982.
23. **Kaplan, H. S. and Tsuchitani, P. J.**, *Cancer in China*, Alan R. Liss, New York, 1978, 115.
24. **Stehr, P. A.**, Vitamin A deficiency as a predisposing factor in the development of stomach cancer, *Diss. Abstr. Int.*, 43, 2863B, 1983.
25. **Hirayama, T.**, A large cohort study on the relationship between diet and selected cancers of the digestive organs, in *Gastrointestinal Cancer: Endogenous Factors*, Banbury Report No. 7, Bruce, W. R., Correa, P., Lipkin, M., Tannenbaum, S. R., and Wilkins, T. C., Eds., Cold Spring Harbor Laboratories, New York, 1981, 409.
26. **Ames, B. N.**, Dietary carcinogens and anticarcinogens. Oxygen radicals in degenerative diseases, *Science*, 221, 1256, 1983.
27. **Wattenberg, L. W.**, Inhibition of neoplasia by minor dietary constituents. *Cancer Res.*, 43, 2448, 1983.
28. **Sato, T., Fukuyama, T., Susuki, T., and Takayanagi, J.**, The relationship between gastric cancer mortality rate and salted food intake in several places in Japan, *Bull. Inst. Publ. Health*, 8, 187, 1959.
29. **Nagai, M., Hashimoto, T., Yanagawa, H., Yokoyama, H., and Minowa, M.**, Relationship of diet to the incidence of esophageal and stomach cancer in Japan, *Nutr. Cancer*, 3, 257, 1982.
30. **Haenszel, W., Kurihara, M., Locke, F. B., Shimuzu, K., and Segi, M.**, Stomach cancer in Japan, *J. Natl. Cancer Inst.*, 56, 265, 1976.
31. **Tuyns, A.**, Sodium chloride and cancer of the digestive tract, *Nutr. Cancer*, 4, 198, 1983.
32. **Joossens, J. V.**, Stroke, stomach cancer and salt. A possible clue to the prevention of hypertension, in *Epidemiology of Arterial Blood Pressure*, Kesteloot, H. and Joossens, J. V., Eds., Martinus Nijhoff Publishers, Boston, 1980, 489.
33. **MacDonald, W. E., Anderson, F. H., and Hashimoto, S.**, Histological effect of certain pickles on the human gastric mucosa, *Can. Med. Assoc. J.*, 96, 1521, 1967.
34. **Sato, T., Fukuyama, T., Urata, G., and Suzuki, T.**, Bleeding in the glandular stomach of mice by feeding highly salted foods and a comment on salted foods in Japan, *Bull. Inst. Publ. Health*, 8, 10, 1959.
35. **Takahashi, M., Kokuho, T., Furukawa, F., Kurokawa, Y., Tatematsu, M., and Hayashi, Y.**, Effect of high salt diet on rat gastric carcinogenesis induced by MNNG, *Gann Monogr.*, 74, 28, 1983.
36. **Schirai, T., Imaida, K., Fukushima, S., Hasegawa, R., Tatematsu, M., and Ito, N.**, Effects of NaCl, Tween 60 and a low dose of N-ethyl-N'-nitro-N-nitrosoguanidine on gastric carcinogenesis of rats given a single does of N-methyl-N'-nitro-N'nitrosoguanidine, *Carcinogenesis*, 12, 1419, 1982.
37. National Academy of Sciences, The Health Effects of Nitrate, Nitrite and N-Nitroso Compounds, National Academy Press, Washington,D. C., 1981, 9-03. 9-17.
38. **Cuello, C., Correa, P., Haenszel, W., Gordillo, G., Brown, C., Archer, M., and Tannenbaum, S. R.**, Gastric cancer in Colombia. I. Cancer risk and suspected environmental agents, *J. Natl. Cancer Inst.*, 57, 1015, 1976.
39. **Hill, M. J., Hawksworth, G., and Tattersall, G.**, Bacteria, nitrosamines and cancer of the stomach, *Br. J. Cancer*, 28, 562, 1973.
40. **Davis, J. M.**, Stomach cancer mortality in Worsop and other Nottinghamshire mining towns, *Br. J. Cancer*, 41, 4389, 1980.
41. **Forman, D., Ab Dabbagh, S., and Doll, R.**, Geographic and social class variations in the U.K. in levels of salivary nitrates and nitrites. A preliminary report, presented at 8th IARC Symposium on N-nitroso Compounds, Banff, 1983, Proceedings in press.
42. **Eisenbrandt, G., Adam, B., Peter, M., Malfertheiner, P., and Schlag, P.**, Formation of nitrite in gastric juice of patients with various gastric disorders after ingestion of a standard dose of nitrate. A possible risk factor in gastric carcinogenesis, presented at 8th IARC Symposium on N-nitroso Compounds, Banff, 1983, Proceedings in press.
43. **Montes, G., Cuello, C., Gordillo, G., Pelon, W., Johnson, W., and Correa, P.**, Mutagenic activity of gastric juice, *Cancer Lett.*, 7, 307, 1979.
44. **Schlag, P., Ulrich, H., Merkle, P., Bockler, R., Peter, M., and Herfarth, C.**, Are nitrite and N-nitroso compounds in gastric juice risk factors for carcinoma of the operated stomach?, *Lancet*, 1, 727, 1980.

45. **Sander, J., Burkle, G., and Schweinsberg, F.,** Induktion malignen tumoren bei ratten durch gleichzeitge verfutterung von nitrit und sek underen aminen. *Z. Krebsforsch.,* 73, 54, 1969.

46. **Weisburger, J. H., Marquardt, H., Hirota, N., Mori, H., and Williams, G. M.,** Induction of cancer of the glandular stomach in rats by extract of nitrite-treated fish, *J. Natl. Cancer Inst.,* 64, 163, 1980.

47. **Correa, P.,** The gastric precancerous process. *Cancer Surv.,* 2, 437, 1983.

48. **Hawksworth, G., Hill, M. J., Gordillo, G., and Cuello, C.,** Possible relationship between nitrates, nitrosamines and gastric cancer in southwest Colombia, in N-nitroso Compounds in the Environment, IARC Scientific Publication No. 9, Lyon, France, 1975, 229.

49. **Tannenbaum, S. R., Moran, D., Falchuk, K. R., Correa, P., and Cuello, C.,** Nitrite stability and nitrosation potential in human gastric juice, *Cancer Lett.,* 14, 131, 1981.

50. **Ruddell, W. S. J., Bone, E. S., Hill, M. J., and Walters, C. J.,** Pathogenesis of gastric cancer in pernicious anemia, *Lancet,* 1, 521, 1978.

51. **Schlag, P., Bockler, R., Meyer, H., and Belohlavek, D.,** Nitrite and N-nitrosocompounds in the operated stomach, in *Gastric Cancer,* Herfarth, Ch. and Schlag, P., Eds., Springer-Verlag, New York, 1979, 120.

52. **Reed, P. I., Haines, K., Smith, P. L. R., House, R. F., and Walters, C. L.,** Gastric juice N-nitrosamines in health and gastrointestinal disease, *Lancet,* 2, 550, 1981.

53. **Piacek-Llanes, B. and Tannenbaum, S. R.,** Formation of an activated N-nitroso compound in nitrite-treated fava beans (Vicia faba), *Carcinogenesis,* 3, 1379, 1982.

54. **Shuker, D. E. G., Tannenbaum, S. R., and Wishnok, J. S.,** N-nitroso bile acid conjugates. I. Synthesis, chemical reactivity and mutagenic activity. *J. Org. Chem.,* 46, 2092, 1981.

55. **Stalsberg, H. and Taksdal, S.,** Stomach cancer following gastric surgery for benign conditions, *Lancet,* 2, 1175, 1971.

56. **Langhans, P., Heger, R. A., Hoberstein, J., and Bunte, H.,** Operation-sequel carcinoma. An experimental study, *Hepato-Gastroenterology,* 28, 34, 1981.

57. **DeFlora, S., Bennicelli, C., Camoirano, A., and Zanachi, P.,** Genotoxicity of nitrosated ranitidine, *Carcinogenesis,* 3, 225, 1983.

58. **Stemmermann, G. N. and Hayashi, T.,** Intestinal metaplasia of the gastric mucosa: a gross and microscopic study of its distribution in various disease states, *J. Natl. Cancer Inst.,* 41, 627, 1968.

59. **Oshima, H. and Bartsch, H.,** Quantitative estimation of endogenous nitrosation in humans by monitoring N-nitrosoproline excreted in the urine, *Cancer Res.,* 41, 3568, 1981.

60. **Newmark, H. L. and Mergens, W. J.,** Blocking nitrosamine formation using ascorbic acid and alpha-tocopherol, in *Gastrointestinal Cancer: Endogenous Factors,* Banbury Report No. 7, Bruce, R., Correa, P., Lipkin, M., Tannenbaum, S. R., and Wilkins, T. D., Eds., Cold Spring Harbor Laboratory, New York, 1981, 285.

61. **Kokatnur, M. G., Murray, M. L., and Correa, P.,** Mutagenic properties of nitrosated spermidine, *Proc. Soc. Exp. Biol. Med.,* 158, 85, 1978.

62. **Kawasaki, H., Morishige, F., Tanaka, H., and Kimoto, E.,** Influence of oral supplementation of ascorbate upon the induction of N-methyl-N'-nitro-N-nitrosoguanidine. *Cancer Lett.,* 16, 57, 1982.

63. **Peto, R., Doll, R., Buckley, J. D., and Sporn, M. B.,** Can dietary beta-carotene materially reduce human cancer rates?, *Nature (London),* 290, 201, 1981.

64. **Sporn, M. B. and Roberts, A. B.,** Role of retinoids in differentiation and carcinogenesis, *Cancer Res.,* 43, 3034, 1983.

Chapter 2

DIET AND CANCER OF THE PANCREAS: EPIDEMIOLOGICAL AND EXPERIMENTAL EVIDENCE

Daniel S. Longnecker and Reginald G. H. Morgan

TABLE OF CONTENTS

I. INTRODUCTION

The pancreas plays a central role in nutrition and metabolism by virtue of the production of digestive enzymes by the exocrine tissue and the secretion of polypeptide hormones by the islets. Although the pancreas is usually thought to be a stable organ with a low rate of cell turnover, it shows rapid change in function and size in response to dietary stimuli. The fact that the ratio of digestive enzymes produced by the pancreas changes in response to long-term changes in the composition of the diet is well established.[1] Altered ratios of exocrine enzymes in pancreatic juice have been reported as soon as 1 to 2 days after a change in diet.[2] Certain diets have been shown to cause hyperplasia of the pancreas and other abnormal diets have caused atrophy. Fasting for longer than 48 hr leads to a fall in DNA synthesis in the rat pancreas and to acinar cell destruction in the mouse,[3] while feeding restores these parameters to normal. Conversely, administration of trypsin inhibitor by gavage or feeding raw soya flour (RSF) leads to an 8- to 10-fold stimulation of DNA synthesis within 24 to 48 hr in the rat and prolonged feeding of RSF leads to pancreatic hypertrophy and hyperplasia. This hyperplasia is maintained as long as the dietary stimulus is continued, but is reversible with withdrawal of RSF from the diet in short-term studies. Similar rapid changes in pancreatic DNA synthesis have been reported following the injection of GI hormones such as cholecystokinin (CCK), gastrin, and secretin. All of these hormones are normally released endogenously in response to dietary stimuli. In view of the ability of the pancreas to respond to dietary stimuli with rapid changes in DNA synthesis and cell turnover, dietary manipulation might be expected to affect pancreatic carcinogenesis.

Carcinomas of the pancreas rank fifth as a cause of cancer deaths in the U.S. More than 90% of these cancers arise in the exocrine portion of the pancreas, and the remainder arise in the islets. The latter are separately designated as islet cell tumors and carcinomas. This discussion will focus on carcinomas of the exocrine pancreas.

Pancreatic carcinomas frequently remain clinically "silent" until late in the disease because of the remote internal location of the pancreas and because the organ possessses a large reserve capacity in regard to both exocrine and endocrine function.[4] Secretion by the exocrine pancreas must be reduced by more than 90% before there is clinical evidence of malabsorption. Carcinomas have usually spread to local or regional lymph nodes before diagnosis, accounting in part for the poor 5-year survival data which are in the range of 1 to 10% in various series. Pain is the most common presenting symptom and carries the implication that there has been retroperitoneal spread involving nerves. The venous drainage of the pancreas is into the portal system, so that metastasis to the liver usually precedes systemic spread. The disease may cause biliary and/or intestinal obstruction, but usually kills because of spread and a high tumor burden. The fact that treatment of advanced disease is usually unsuccessful imposes a special need to recognize the causes of carcinoma of the pancreas in the hope that steps may be taken to reduce the incidence.

Several recently reported epidemiological studies have evaluated the association of various risk factors with pancreatic carcinoma. Few useful clues have been provided by such studies. Cigarette smoking appears to be the best established life style risk factor.[5,6] Among dietary and nutritional factors, high dietary fat,[7] high meat consumption,[8] and coffee drinking[9] have been implicated. The validity of each of these factors has been questioned because of the failure to detect an association in other studies as reflected in recent reviews.[5,6]

Animal models for pancreatic carcinogenesis have been developed in the past decade and utilized in several studies that have shown that dietary factors may influence the experimental induction of pancreatic carcinomas. The effects of nutritional factors on the pancreas and on experimental carcinogenesis are more easily recognized in the laboratory where attention can be focused on a single factor under controlled conditions than in epidemiologic studies of human populations.

II. PANCREATIC CARCINOGENESIS IN ANIMAL MODELS

More than a dozen chemicals have induced pancreatic carcinomas in laboratory animals.[10] Two models have emerged as the best characterized and most widely studied. These are the treatment of hamsters with *N*-nitrosobis(2-oxopropyl)amine (BOP)[11] and the treatment of rats with azaserine.[12] Earlier reviews should be consulted for detailed descriptions and comparison of these models, but a few comments are in order here. Carcinomas can be induced in both models with a single dose of carcinogen although it is sometimes preferable to use a series of 2 to 5 weekly injections to achieve the desired incidence. Thus it is possible, in both models, to complete carcinogen treatment in a short interval so that diets can be manipulated separately during, or following, carcinogen treatment.

The dominant histologic type of the carcinomas differs in the two models. In the rat, most carcinomas show evidence of acinar cell differentiation,[10,12] and there is good evidence that all carcinomas, even those that are undifferentiated or include duct-like areas, have arisen from acinar cells. The dominant histologic type of carcinoma in the hamster is duct-like in appearance although a few acinar cell carcinomas have been observed.[10] The nature of early lesions and the results of biochemical studies indicate that BOP affects both acinar and ductal cells. The cell of origin of carcinomas in this model remains a matter of some controversy. Various investigators feel that carcinomas arise exclusively from ductal or ductular cells,[13] exclusively from acinar cells,[14] or from both.[15] Pour and co-workers have proposed that the hamster provides a better model than the rat for human pancreatic cancer, because the majority of human carcinomas are of ductal histologic type.[16] Thus, the resolution of the cell of origin in the hamster may have special implications for the human.

In the rat, it appears to be largely the acinar cell that responds to trophic stimuli. Although the acinar cell is highly differentiated for protein synthesis and secretion, such cells (or a subset) are capable of cell division. Autoradiographic studies of cell turnover in the rat pancreas show labeling of acinar cells. Division of acinar cells is occasionally seen in the normal adult pancreas and is markedly increased in regenerating pancreas.[1,15] In the rat fed RSF or injected with CCK, and in the regenerating rat pancreas, increased cell division also occurs in duct and centroacinar cells. Although these cells represent only a very small fraction of the total cell mass, increased labeling in this compartment has been reported in some studies to precede labeling in the acinar cell compartment, leading to the suggestion that some cells with centroacinar morphology might represent a population of stem cells which responds to the trophic stimulus by division and differentiation into duct, acinar, or islet cells.[17] The relative importance of new cells derived from such putative stem cells as opposed to division of differentiated cell types has not been determined.

A. Effect of Caloric Restriction

Restricted consumption of diet has been shown to reduce the incidence of pancreatic neoplasms in the azaserine/rat model of pancreatic carcinogenesis when calorie-restricted and random-fed groups of carcinogen-treated animals were compared.[18] This effect was dramatic in azaserine-treated rats in an experiment in which the incidence of neoplasms was reduced from 4/17 in the *ad lib*-fed group to 0/10 in a group fed 90% of the amount of diet consumed by the former group during the 6-week period of exposure to carcinogen. Both groups were fed *ad libitum* during the 7.5-month postinitiation phase of this experiment. In a third group, the diet was restricted only during the postinitiation phase and 1/9 rats developed a pancreatic neoplasm (not a significant difference from the *ad lib*-fed control group). Thus, it appeared that the effect of restricted caloric intake was exerted primarily during the period of exposure to carcinogen. Similar effects of caloric restriction on carcinogenesis in other tissues have been reported.[19]

Two possible explanations for the effect of caloric restriction on carcinogenesis in the pancreas include (1) alteration in xenobiotic metabolism and (2) reduced trophic stimulation to the pancreas. It has been reported that levels of drug metabolizing enzymes are altered in animals with restricted food intake.[20] It is also reasonable to assume that less CCK is released from the intestine in animals with reduced food consumption, and therefore that the rate of cell division in the pancreas might be reduced.

B. Effect of Purified Vs. Chow Diets

Comparison of the incidence of pancreatic neoplasms in groups of azaserine-treated rats fed chow or defined purified diets has suggested that some component of chow inhibits the progression of carcinogen-induced lesions. In one experiment, 12/17 rats fed the AIN-76A diet developed pancreatic neoplasms while the incidence was 4/19 among chow-fed rats.[12] Subsequent experiments indicated that this effect of chow was exerted primarily during the postinitiation phase of carcinogenesis, i.e., after exposure to carcinogen was completed.[12] The reason for this difference in rate of progression of carcinogen-induced lesions in the pancreas is not known, but the possibility that some unidentified absorbable component of chow serves as a chemopreventive agent is attractive.

C. Effect of RSF Feeding in the Rat

Although soya flour is rich in high quality protein (approximately 40%) and fat (20%), in many species, particularly the chicken and the rat, RSF feeding causes growth depression associated with marked pancreatic enlargement.[21-25] This effect is largely due to a heat-labile component as discussed below since adequate heating of the flour results in a product with little effect on pancreatic growth and no deleterious effects on overall growth.[23,25] In the Wistar rat the pancreatic enlargement produced by RSF feeding is initially due to uniform hypertrophy and hyperplasia of the acinar cells, but after about 24 weeks a marked increase in the incidence of multifocal hyperplasia producing atypical acinar cell foci and nodules is seen. Some of the nodules progress to become adenomas in 80 to 100% of animals after about 60 weeks and to carcinoma in more than 10% of animals after more than 90 weeks of RSF feeding.[26-28] The lesions at all stages are morphologically similar to corresponding neoplasms in chow-fed animals. As indicated above, heated soya flour (HSF) feeding does not cause pancreatic enlargement, but nevertheless a small increase in the incidence and the rate of progression of atypical acinar cell foci does occur.[27,28]

RSF thus promotes spontaneous neoplastic change in the pancreas in the absence of exogenous carcinogen. In addition, it potentiates the action of known pancreatic carcinogens in rat. The amino acid derivative azaserine causes a high incidence of acinar cell-derived carcinomas in the rat, with progression of the lesions through atypical acinar cell foci to adenocarcinoma as in the spontaneous tumors.[12,29] RSF greatly potentiates the incidence and the rate of progression of these lesions.[27,30] The nitrosamine N-nitrosobis(2-hydroxy-propyl)amine (BHP) causes pancreatic carcinoma apparently derived from ductal cells in high yield in the hamster[31] but has little carcinogenic effect on the pancreas in the rat fed chow[31-33] or HSF.[33] In the rat fed RSF, however, BHP produced many acinar cell neoplasms, progressing from atypical acinar cell foci and nodules through adenoma to acinar cell-derived carcinoma. This potentiation occurs despite a significant depression of the RSF-induced pancreatic hypertrophy by the carcinogen.[33]

In addition to its effect on the pancreas, BHP produced a high incidence of tumors in other organs in both the hamster and the rat. RSF feeding did not affect the incidence of tumors in any organ other than the pancreas, emphasizing the specificity of the effects of the RSF feeding for the pancreas.[33]

The studies discussed above on the carcinogenic effects of RSF and the promotion of pancreatic carcinogens by this diet all used a diet of 100% RSF supplemented with vitamins and minerals. A detailed investigation of the dose-response relationship between RSF and pancreatic carcinogenesis has not yet been performed, but in a study in which diets consisting solely of varying mixtures of RSF and HSF were fed, both spontaneous and azaserine-induced neoplastic changes in the pancreas were significantly increased at the lowest RSF ratio tested, 25% RSF/75% HSF.[28] As will be discussed later, the most important component of the RSF diets as far as nodule production is concerned is probably trypsin inhibitor. A 25% RSF diet as used in these studies would supply approximately 1000 mg of trypsin inhibitor per 100-g diet. Studies of the effects of lower levels of RSF on pancreatic neoplasia have not yet been published, but an extensive investigation of a number of soya products in varying doses is in progress with funding from the U.S. Department of Agriculture. Details of this study are not yet available, but a direct relationship between the trypsin inhibitor (TI) content of the diet and the incidence of pancreatic neoplasms in the rat has been found, with a significant increase, compared with control diets, in the number of nodules at trypsin inhibitor levels of 337 mg TI per 100-g diet. No increase occurred at 215 mg or less TI per 100-g diet. The total protein content of the diet affected these results, nodule incidence being greater at 20% total protein than at 10 or 30% protein.[96] It appears, then, that quite low levels of trypsin inhibitor may accelerate neoplastic changes in the pancreas.

The critical length of time of exposure to RSF necessary to produce neoplastic changes in the pancreas is unknown. If the diet is changed back to chow or HSF after 4 weeks RSF feeding, complete and rapid involution of the pancreas occurs within 1 week.[34] After continuous exposure, microscopic atypical acinar cell foci are first detectable in low yield after 16 to 20 weeks, and microscopic nodules after 24 weeks.[30] If RSF feeding is limited to a period of 12 weeks, animals killed a year or more later show no increased incidence of neoplasms.[28] No systematic study of the effect of changing back to a chow diet after intervals longer than this have been reported, but unpublished observations suggest that continous administration of RSF for less than 6 months is reversible, while RSF feeding for longer than 6 months results in irreversible nodule formation and progression to pancreatic carcinoma, despite later change to nonsoya flour-containing diets.[97]

Two studies suggest that prolonged intermittent exposure to RSF potentiates spontaneous and azaserine-induced carcinogenesis at least as well as continous exposure. Thus, a group of 5 rats were fed RSF for 2 days (over the weekend) with rat chow for the remaining 5 days each week for 90 weeks; 3 of the 5 animals developed pancreatic cancer after 60 weeks while the remainder had grossly adenomatous pancreases when killed at the end of the study.[28] In the second study,[35] rats were fed a diet alternating weekly between RSF and rat chow, with azaserine or saline injections (5 mg/kg twice weekly during the week on RSF for a total of 24 injections). When the animals were killed while eating RSF after 24 to 36 weeks, the incidence of foci and nodules was not significantly different from that seen in corresponding azaserine or saline injected animals fed RSF continuously. At earlier stages (16 and 24 weeks) the incidence of nodules fluctuated significantly with changes in diet, being higher while the animals were fed RSF and lower while on chow. At all times studied (up to 65 weeks), mitotic activity in the nodules was significantly increased when the animals were killed during the week of RSF feeding compared with the week of chow feeding. It appears, therefore, that even quite advanced nodules respond vigorously to the growth stimulus produced by RSF feeding.

1. Mechanism of Action of RSF on the Pancreas

Most of the pancreatic enlargement seen in animals fed RSF is probably due to the high content of trypsin inhibitors in RSF. These inhibitors are largely destroyed by moist heat, and HSF with minimal inhibitor activity has little effect on pancreatic growth.[23,25] The inhibitor content varies between strains of soya beans and there is a good correlation between the inhibitor content of different preparations of soya flour and pancreatic size in rats fed these preparations.[36] In addition, numerous other studies have demonstrated pancreatic enlargement in the rat fed trypsin inhibitors from other legumes[37] including peanuts,[38] egg white,[37,39] bovine lung,[40] and synthetic inhibitors.[41]

A mechanism to link trypsin inhibitor feeding to pancreatic growth was suggested by Green and co-workers.[42-44] In a series of studies, convincing evidence was presented that trypsin in the intestinal lumen inhibited the release of CCK. When trypsin levels were lowered by the presence of undigested protein or, more potently, trypsin inhibitor, in the gut lumen, CCK release was stimulated. Most of these studies used increased pancreatic secretion as an index of CCK release, but since CCK is a trophic hormone for the pancreas, the extension of the theory to explain pancreatic hypertrophy in trypsin inhibitor-fed animals was obvious.

No serious objections to this theory have been raised but it has proven very difficult to convincingly demonstrate that CCK release does in fact occur after trypsin inhibitor feeding. Nevertheless, a loss of CCK-like activity from the gut wall, as measured by gall bladder bioassay, and an increase in circulating pancreozymin-like activity as measured by decreased pancreatic and increased intestinal amylase activity occurred within 30 min of a gavage of a soyabean trypsin inhibitor preparation.[45]

More direct evidence for or against CCK being the mediator of the pancreatic enlargement produced by RSF feeding should be forthcoming as reliable radioimmunoassays for CCK are more widely available. Abstracts reporting raised plasma CCK levels in rats 21 days after starting inhibitor feeding,[46] and raised CCK levels in rats with pancreatic acinar cell atrophy (in which intestinal trypsin levels are reduced)[47] have been published, but details of these studies are not yet available.

The effect of RSF in promoting neoplastic changes in the pancreas is probably linked to the hyperplasia produced by CCK since pancreatic growth produced by other means also potentiates pancreatic carcinogenesis. Thus, partial pancreatectomy[48] and ethionine-induced pancreatitis[49] both lead to increased pancreatic growth (regeneration) and both potentiate azaserine in the rat. Partial pancreatectomy also potentiates nitrosamine carcinogenesis in the hamster.[50] Just how this stimulation of growth leads to neoplasia, though, is not clear. It is not likely that spontaneous neoplastic foci simply develop faster as growth of the nodules is stimulated since this should lead to a more rapid progression, but not an increased number of nodules. The fact that nodule number does increase indicates that additional initiated cells progress to become foci and nodules, possibly arising in cells that would not proliferate with only physiologic growth stimuli. Probably, in addition, the dividing "normal" cells involved in the hyperplasia are more sensitive to carcinogens[51,52] leading to an increased incidence of initiated foci as well as their rate of growth.

In addition to trypsin inhibitor, other factors in RSF may have a role in promoting pancreatic growth. As was mentioned above, in rats fed HSF, a small but readily detected increase in spontaneous and azaserine-induced neoplasms occurs. Almost certainly a large part of this effect is due to the high unsaturated fat content of this diet, as is discussed later. Furthermore, soya flour contains a number of other toxic components[53] and the role of these in pancreatic growth promotion and neoplasia has not been assessed.

2. Other Dietary Sources of Trypsin Inhibitors and Stimulants of Pancreatic Growth

Many legumes and other vegetables contain trypsin inhibitors[54] and might be expected to promote pancreatic neoplasia. No studies on the effects of trypsin inhibitors from other foods on pancreatic neoplasms have been published, but the U.S. Agriculture Department study in progress mentioned earlier includes an investigation of the long-term effect of feeding diets containing purified potato trypsin inhibitor. Preliminary results suggest that potato inhibitor is as effective as soyabean inhibitor in promoting pancreatic growth, and therefore would presumably enhance nodule development when given at the same dose level.

Other trophic hormones for the pancreas include secretin, gastrin,[55] and epidermal growth factor.[56] No studies have looked at the effect of these hormones in pancreatic neoplasia, but any dietary manipulation or drug therapy which leads to prolonged changes in blood levels of these hormones might also be capable of modulating pancreatic carcinogenesis.

D. The Effect of RSF in Other Species

The effects of prolonged RSF feeding on pancreatic neoplasia in species other than the rat have not been reported. Although soyabean trypsin inhibitor is active against trypsin from a wide variety of species,[57,58] pancreatic hypertrophy after RSF feeding does not occur in some animals. For example the dog does not respond,[59,60] and therefore it seems unlikely that RSF would potentiate pancreatic carcinogenesis in this species. It has been reported that the hamster pancreas responds to RSF feeding[61] but the significance of this finding is obscure. In this study, pancreatic weight relative to body weight was increased in hamsters fed a trypsin inhibitor preparation, but inhibitor feeding caused significant depression of growth in these hamsters, and absolute pancreatic weight was the same in inhibitor-fed and control groups (287 vs. 288 mg). The effect of RSF feeding on the hamster pancreas may therefore not be as marked as that seen in the rat but nevertheless a study of the effects of RSF feeding on nitrosamine-induced pancreatic neoplasms in this species would be particularly interesting.

It is important to know whether the human pancreas responds to dietary trypsin inhibitor. Obviously direct information on the effect of RSF on pancreatic growth in the human will be virtually impossible to obtain, but studies on pancreatic secretion after trypsin inhibitor feeding or diversion of pancreatic juice both support[62,63] and deny[64,65] the existence of a negative feedback loop involving trypsin and CCK in the human. It is possible that studies of RSF feeding in the primate and the measurement of plasma CCK levels in the human fed RSF or trypsin inhibitor will indicate whether trypsin inhibitor feeding is likely to promote pancreatic neoplasia in man. Whatever the answer, it seems unlikely that the normal cooked human diet contains significant levels of trypsin inhibitor. Doell et al.[66] have estimated that the average cooked British diet contains trypsin inhibitor activity sufficient to inhibit 330 mg of bovine trypsin per day. Half of this activity was present in eggs and potatoes, the trypsin inhibitors of which are said to be inactive against human trypsin, so the inhibitor activity against human trypsin is probably less than 150 mg/day. It is possible, however, that increased use of soyabean products could increase this value.

E. Effect of High-Fat Diets

Epidemiologic studies have shown a general correlation of national levels of dietary fat consumption with the incidence of carcinoma of the breast[67] and of the pancreas.[7] In experimental studies, high-fat diets have been shown to enhance carcinogenesis in the breast, colon, and skin.[67] This has stimulated similar studies with experimental models of pancreatic carcinogenesis. Enhanced development of pancreatic carcinomas has been demonstrated in azaserine-treated rats that were fed diets containing 20% corn oil or 20% safflower oil and in N-nitrosobis(2-oxopropyl)amine-treated hamsters fed a 20% corn oil diet during the post-initiation phase of carcinogenesis.[18,68,69] It is of interest that the incidence of pancreatic neoplasms was similar in a group of rats fed a diet containing 18% coconut oil plus 2% corn oil and in a group fed a diet containing 5% corn oil,[18] and that the incidence in both of these groups was significantly lower than in groups fed diets containing 20% unsaturated fat. This observation suggests that unsaturated essential fatty acids may play a critical role in the progression of carcinogen-induced lesions. In the study in the hamster model, three dietary levels of corn oil, 4.25, 9, and 20%, were compared.[68] Both incidence and the average number of carcinomas per pancreas increased with the dietary fat level. The incidence was 16, 20, and 34% and the multiplicity was 1.3, 1.9, and 3.0 cancers per tumor-bearing animal at the three fat levels. These data suggest that there may be a nearly linear dose response in enhancement of carcinogenesis with increasing levels of unsaturated fat although further data are required to verify this point.

Enhancement of growth of carcinogen-induced foci and nodules in the pancreas of rats fed high-fat diets has been demonstrated in experiments of 4-months duration,[70] but less is known about the reversibility of the effect of unsaturated fats on pancreatic carcinogenesis than is the case for RSF. To date, there have been few studies of the effect of feeding high-fat diets for limited periods. Roebuck et al. reported that feeding a 20% corn oil diet only during the period of exposure to azaserine did not enhance the number of neoplasms developing during a 9-month experiment,[18] whereas feeding the high-fat diet beginning 1 week after exposure to azaserine had been completed did enhance the development of neoplasms. We are not aware of experiments in which feeding of the high-fat diets has been delayed until long after carcinogen treatment.

Because of the reports that corn oil enhances the progression of carcinogen-induced lesions, a question has arisen regarding the effect of high corn oil intake on the pancreas of non-carcinogen-treated animals. A retrospective review of the incidence of acinar cell nodules and adenomas among noncarcinogen-treated control rats involved in bioassay studies under the auspices of the National Toxicology Program has been completed. Most such rats were given corn oil by gavage at the rate of 5 mℓ/kg, 5 days/week in lifetime studies. Male rats subjected to this regime developed a low, but significantly elevated incidence of nodules and adenomas in comparision with nongavage controls.[71] In this study, the number of lesions per pancreas was seldom more than one. This result is consistent with enhanced growth of "spontaneously" occurring foci of atypical acinar cells since such lesions have been reported in as many as 75% of aged rats.[12]

The mechanism of this effect of high dietary levels of unsaturated fat is unknown. The pancreas has been shown to increase the synthesis and secretion of lipase in response to high dietary levels of fat. This mechanism appears at least in part to be a response to circulating levels of fatty acids and monoglycerides,[1] but there is no clear evidence that high-fat diets cause pancreatic hyperplasia. The fact that high-fat diets have a similar effect on carcinogenesis in several diverse organs suggests that there may be an effect on pancreatic parenchymal cells that is not closely linked to unique aspects of pancreatic physiology.

F. Effect of Altered Levels of Dietary Protein

The impact of protein deprivation[72] and single amino acid deficiency[73] on the pancreas has been demonstrated in experimental animals. Pancreatic atrophy has been described in rats fed diets that were deficient in histidine, isoleucine, leucine, lysine, phenylalanine, tryptophan, threonine, valine, or total protein, but atrophy did not develop with deficiencies of methionine or arginine.[73] Similar changes were noted in rats that were force-fed several plant proteins of poor quality, but not in rats that were allowed to eat these diets *ad libitum*.[74] In the aggregate, these studies demonstrate the sensitivity of the acinar cell to protein deficiency and amino acid imbalances. To a large degree, such atrophy is reversible with correction of the dietary deficiency, i.e., the pancreas can regenerate. As noted above, regeneration of the pancreas following partial pancreatectomy or ethionine-induced atrophy has enhanced the effectiveness of experimental carcinogens. Thus, it seems reasonable that cycles of atrophy and regeneration in response to dietary changes might sensitize the pancreas to the effect of carcinogens. On the other hand, feeding diets with a low protein content has been shown to cause a decrease in levels of certain drug-metabolizing enzymes that might be involved in carcinogen activation.[20,75]

Experimental studies have evaluated the effect of varying levels of dietary protein. Roebuck et al. reported a similar incidence of pancreatic neoplasms in azaserine-treated rats fed diets containing 11 and 20% casein whereas a group fed 50% casein (with reduced dietary carbohydrate) developed fewer neoplasms.[76] The incidences were 8/9, 12/17, and 5/18, respectively, in these three groups. The reason for the reduced incidence of pancreatic adenomas and carcinomas in the last group is not known with certainty, but could reflect a voluntary reduction in caloric intake since the calculated caloric consumption was reported to be 11% lower in the high protein-fed group. The final mean body weight of this group was 87% that of the 20% casein-fed group. Pour et al. have reported a reduced incidence of pancreatic neoplasms among hamsters that were fed a protein-free diet for 28 days immediately before, during, or following a single BOP injection.[77] Pour and Birt have also reported a reduced incidence of pancreatic carcinomas in female hamsters fed a diet containing 9% protein and an enhanced incidence in females fed a 36% protein diet following a single BOP injection in comparison with a control group fed 18% protein.[78] These differences from control were not noted in male hamster groups.

In these studies, the protein source was nutritionally adequate (casein), and the diets have been fed continuously rather than cyclicly. Thus, the possibility that cyclic feeding of protein-deficient diets might influence pancreatic carcinogenesis has not been evaluated, and results from studies in animal models show no consistent trend to date in regard to effects of altered protein consumption on pancreatic carcinogenesis.

The interaction of dietary fat and protein in modulation of pancreatic carcinogenesis has been examined in one study done in the hamster model.[69] The result was generally consistent with studies cited above and indicated that BOP-treated hamsters fed diets high in both protein and fat developed more pancreatic carcinomas than hamsters fed diets low in both protein and fat. The low fat diets contained 4.25% corn oil, i.e., about the level of fat provided by chow diets and the standard AIN-76A diet. The high-fat diets contained 20% corn oil. Low and high protein levels were 8.5 to 10.2% and 34 to 41% casein, respectively, in isocaloric diets. The incidence of pancreatic carcinoma was similar in hamsters fed diets that were low in both fat and protein and in hamsters fed a low-fat, high protein diet during the postinitiation period, i.e., following carcinogen treatment. This observation may help to explain the apparent disagreement between studies done in rats and hamsters, since in rats the high protein level was tested only in a diet that contained 5% corn oil.[76] The studies in the hamster model support the contention that there is an interaction between dietary levels of fat and protein that modifies the response of the pancreas to carcinogens.[69]

G. Effect of Altered Selenium Intake

Recent studies indicate that added dietary selenium inhibits the progression of azaserine-induced foci and nodules in rats.[79] Selenium was added to the diets as sodium selenite, a form that has been effective against the experimental induction of carcinoma of the breast.[80] Although the mechanism of this anticarcinogenic effect of selenium is unknown, effectiveness has been demonstrated during both initiation and postinitiation phases of carcinogenesis.[80] It is of interest that Lawson and Birt have reported data indicating that feeding 5 ppm of sodium selenite to hamsters increased parameters of DNA repair in pancreas after injection of 20 mg BOP per kilogram.[81] Evidence of enhanced DNA repair in rat colon was also reported with feeding of 2 ppm sodium selenite.[82]

Pancreatic atrophy and fibrosis has been described in chicks maintained on a selenium-deficient diet.[83] Selenomethionine was highly effective in preventing such pancreatic atrophy, presumably because of its great affinity for the pancreas. A comparison of the effectiveness of sodium selenite with added dietary selenomethionine in the animal models for pancreatic carcinogenesis would be of interest because of the affinity of the pancreas for the latter compound.

H. Effect of Retinoids on Experimental Pancreatic Carcinogenesis

Squamous metaplasia of the interlobular pancreatic ductal epithelium has been described in rats maintained on a vitamin A-free diet for several months,[84] suggesting that vitamin A is required to maintain normal epithelial differentiation in the pancreas as in other sites. This provided a rationale for evaluating the effect of enhanced dietary retinoid intake on pancreatic carcinogenesis in the animal models.

Three long-term experiments in azaserine-treated rats have all provided evidence that addition of high levels of several retinoids inhibited the progression of neoplastic lesions in the pancreas.[12,85,86] For example, the incidence of pancreatic carcinomas was 46% among a group of azaserine-treated male Lewis rats fed chow to which only a solvent mixture (vehicle) was added. In contrast, comparable groups fed chow supplemented with 1 or 2 mmol/kg of N-2-hydroxyethylretinamide (2-HER) for 1 year following completion of carcinogen treatment developed incidences of 8 and 4%, respectively. The higher dose level was toxic, but the rats fed the lower level showed only slight growth depression and weighed 98% as much as the control group at the end of the experiment. 2-HER is one of the least toxic of nine retinoids that we have studied and was one of the most effective in regard to inhibiting the progression of azaserine-induced lesions. Other effective retinoids include retinyl acetate, retinylidene dimedone, N-4-propionyloxyphenylretinamide, N-(2,3-dihydroxypropyl)retinamide, and N-(4-pivaloyloxyphenyl)retinamide.

Similar studies have been done using the BOP/hamster model for pancreatic carcinogenesis with less conclusive results. We have seen a reduced incidence of pancreatic carcinomas in two of four retinoid treated groups in one study[98] and failed to see a significant reduction in the incidence of carcinomas in a second study in which four retinoids were evaluated using lower dietary levels of retinoid.[87] In aggregate, our experience in the hamster model suggests that some retinoids may be effective chemopreventive agents for pancreas. In contrast, Birt et al. have reported an enhanced yield of pancreatic carcinomas in hamsters that were fed diets supplemented with retinoids at the level of 0.4 to 1.0 mmol/kg of diet following a single injection of 40 mg BOP per kilogram body weight.[88] It is of interest that these retinoid levels were similar to those that we used in the study in which we observed a lower incidence of carcinomas in some of the groups of hamsters using a BOP dose of 20 mg/kg. The enhancement of carcinoma incidence was not observed by Birt when the high levels of retinoid were fed following a lower dose of BOP (10 mg/kg). Previous studies from the same group had not shown such enhancement when lower doses of retinoids were fed (0.05 to 0.2 mmol/kg of diet).[89]

Hamsters tolerate retinoids less well than rats, requiring that lower dietary levels be used in long-term studies. This provides one possible explanation for the failure of retinoids to inhibit carcinogenesis in several studies with the hamster model, but provides no explanation for unexpected increase in carcinomas observed by Birt et al.[88] The effect of dietary retinoid supplements on carcinogenesis in the hamster requires further evaluation, especially the suggestion that some retinoids might promote pancreatic carcinogenesis in the hamster under selected conditions.

I. Miscellaneous Factors of Possible Relevance

Atrophy of acinar tissue and its replacement by fat has been reported in rats that were maintained on copper-free diets for prolonged periods. The rate of development of clinically significant copper deficiency was accelerated by feeding a chelating agent, penicillamine, to speed the depletion of tissue stores.[90] Islet and ductal tissue were little affected, and the atrophic changes were not accompanied by significant inflammation or fibrosis. Studies of pancreatic regeneration after copper deficiency have not been reported, and the effect of copper deficiency on pancreatic carcinogenesis is unknown. The studies cited above suggest that regeneration following copper deficiency could modify the sensitivity of experimental animals to pancreatic carcinogens, or that copper deficiency could alter the rate of progression of carcinogen-induced lesions in the pancreas.

It has long been known that ethionine is toxic for the pancreas in rats. Injection of large doses (1 g/kg) induced pancreatitis, while injecting or feeding lower levels caused severe, though reversible, pancreatic atrophy.[91] There was a conspicuous degeneration of acinar cells in such rats, and there was prompt regeneration when ethionine was stopped and methionine was supplemented. As noted above, the pancreas of both rats and hamsters has been shown to be highly susceptible to the effect of carcinogens during regeneration.[15,48,49] While this toxic effect of ethionine may be outside the realm of nutrition in the traditional sense, Lombardi's studies of the induction of acute hemorrhagic pancreatitis with ethionine in mice implicates choline deficiency as an important factor in predisposing the pancreas to this injury.[91] Shinozuka has studied the effect of choline deficiency on pancreatic carcinogenesis and reported that there were fewer carcinogen-induced lesions in the pancreas of azaserine-treated rats fed deficient diets than in rats fed a choline-supplemented diet.[92] This study was 6 months in duration and no carcinomas were observed in either group, but the observation suggests that choline-deficiency could alter the process of carcinogenesis in the pancreas in the azaserine-rat model.

Azaserine appears to be metabolically activated by a pyridoxal-dependent enzyme system that is distinct from the cytochrome oxidase system.[93] Preliminary studies have shown that pyridoxal-deficient rats developed fewer AACN than normal animals in response to a dose of azaserine in a study of 4-months duration, and that pyridoxal-deficient rats showed less evidence of DNA damage in the pancreas 1 hr after azaserine injection than normal animals. This model demonstrates that nutritional factors can influence carcinogenesis in a highly specific manner that may relate only to a single carcinogen.

The chronic pancreatitis associated with chronic alcoholism may more correctly be classed as an example of toxic cellular injury than as a nutritional disease. There has been speculation that this disease might predispose to the development of carcinoma because of cyclic injury and regeneration in the pancreas. Alcohol consumption has been evaluated as a risk factor for carcinoma of the pancreas by several investigators with differing conclusions. The net conclusion from these studies seems more negative than positive in regard to the association.[5,6]

Studies in animal models reported to date have shown a reduced rather than an increased incidence of pancreatic carcinomas in BOP-treated hamsters fed a high level of ethanol,[94] and no effect in a second experiment using a lower level of ethanol intake.[95]

III. CONCLUDING REMARKS

The studies cited above provide convincing experimental evidence that carcinogenesis in the exocrine pancreas can be modulated by diet. Some factors that have been implicated by epidemiologic studies remain to be evaluated in animal models, e.g., coffee consumption. Conversely, the significance of certain factors implicated by studies in animals remains to be evaluated in the human, e.g., the effect of dietary trypsin inhibitors. In the case of the effect of dietary fat, available epidemiologic and experimental data is in reasonable agreement. In this case it will be important to ascertain the mechanism of the enhancing effect of dietary lipid on pancreatic carcinogenesis so that rational intervention can be planned. A continuing dialogue between epidemiologists and experimentalists will be required to complete the assessment of the role of dietary factors in pancreatic carcinogenesis.

ACKNOWLEDGMENTS

The authors thank Elna T.Kuhlmann for editorial assistance and M. B. Glynn for typing. The contributions of colleagues and students to aspects of this work is reflected in the authorship of papers that we have cited from our laboratories.

REFERENCES

1. **Solomon, T. E.,** Regulation of exocrine pancreatic cell proliferation and enzyme synthesis, in *Physiology of the Gastrointestinal Tract,* Vol 2, Johnson, L. R., Ed., Raven Press, New York, 1981, 873.
2. **Christophe, J., Camus, J., Deschodt-Lanckman, M., Rathé, J., Robberecht, P., Vandermeers-Piret, M. C., and Vandermeers, A.,** Factors regulating biosynthesis, intracellular transport and secretion of amylase and lipase in the rat exocrine pancreas, *Horm. Metab. Res.,* 3, 393, 1971.
3. **Nevalainen, T. J. and Janigan, D. T.,** Degeneration of mouse pancreatic acinar cells during fasting, *Virchows Arch. B,* 15, 107, 1974.
4. **Longnecker, D. S.,** Pathology and pathogenesis of the diseases of the pancreas, *Am. J. Pathol.,* 107, 99, 1982.
5. **Mack, T. M.,** Pancreas, in *Cancer Epidemiology and Prevention,* Schottenfeld, D. and Fraumeni, J. F., Jr., Eds., W. B. Saunders, Philadelphia, 1982, 638.
6. **MacMahon, B.,** Risk factors for cancer of the pancreas, *Cancer,* 50, 2676, 1982.
7. **Wynder, E. L.,** An epidemiological evaluation of the causes of cancer of the pancreas, *Cancer Res.,* 35, 2228, 1975.
8. **Hirayama, T.,** A large-scale cohort study on the relationship between diet and selected cancers of digestive organs, in Banbury Report No. 7 Gastrointestestinal Cancer: Endogenous Factors, Bruce, W. R., Correa, P., Lipkin, M., Tannenbaum, S. R., and Wilkins, T. D., Eds., Cold Spring Harbor Laboratory, New York, 1981.
9. **MacMahon, B., Yen, S., Trichopoulos, D., Warren, K., and Nardi, G.,** Coffee and cancer of the pancreas, *N. Engl. J. Med.,* 304, 630, 1981.
10. **Longnecker, D. S., Wiebkin, P., Schaeffer, B. K., and Roebuck, B. D.,** Experimental carcinogenesis in the pancreas, in *International Review Experimental Pathology,* Vol. 26, Richter, G. W. and Epstein, M. A., Eds., Academic Press, New York, 1984, 177.
11. **Pour, P. M., Runge, R. G., Birt, D., Gingell, R., Lawson, T., Nagel, D., Wallacave, L., and Salmasi, S. Z.,** Current knowledge of pancreatic carcinogenesis in the hamster and its relevance to the human disease, *Cancer,* 47, 1573, 1981.
12. **Longnecker, D. S., Roebuck, B. D., Yager, J. D., Jr., Lilja, H. S., and Siegmund, B. T.,** Pancreatic carcinoma in azaserine-treated rats: induction, classification, and dietary modulation of incidence, *Cancer,* 47, 1562, 1981.
13. **Pour, P., Althoff, J., and Takahashi, M.,** Early lesions of pancreatic ductal carcinoma in the hamster model, *Am. J. Pathol.,* 88, 291, 1977.
14. **Flaks, B., Moore, M. A., and Flaks, A.,** Ultrastructural analysis of pancreatic carcinogenesis. V. Changes in differentiation of acinar cells during chronic treatment with N-nitrosobis(2-hydroxypropyl)amine, *Carcinogenesis,* 3, 485, 1982.
15. **Scarpelli, D. G., Rao, M. S., and Subbarao, V.,** Augmentation of carcinogenesis by N-nitrosobis(2-oxopropyl)amine administered during S phase of the cell cycle in regenerating hamster pancreas, *Cancer Res.,* 43, 611, 1983.
16. **Cubilla, A. L. and Fitzgerald, P. J.,** Cancer (non-endocrine) of the pancreas. A suggested classification, in *Monographs in Pathology #21, The Pancreas,* Fitzgerald, P. J. and Morrison, A. B., Eds., Williams & Wilkins, Baltimore, Md., 1980, 82.
17. **Adler, G., Hupp, T., and Kern, H. F.,** Course and spontaneous regression of acute pancreatitis in the rat, *Virchows Arch. A,* 382, 31, 1979.
18. **Roebuck, B. D., Yager, J. D., Jr., Longnecker, D. S., and Wilpone, S. A.,** Promotion by unsaturated fat of azaserine-induced pancreatic carcinogenesis in the rat, *Cancer Res.,* 41, 3961, 1981.
19. **Tannenbaum, A.,** Effects of varying caloric intake upon tumor incidence and tumor growth, *Ann. N.Y. Acad. Sci.,* 49, 5, 1947.
20. **Campbell, T. C. and Hayes, J. R.,** Role of nutrition in the drug-metabolizing enzyme system, *Pharmacol. Rev.,* 26, 171, 1974.
21. **Chernick, S. S., Lepkovsky, S., and Chaikoff, I. L.,** A dietary factor regulating the enzyme content of the pancreas: changes induced in size and proteolytic activity of the chick pancreas by the ingestion of raw soy-bean meal, *Am. J. Physiol.,* 155, 33, 1948.
22. **Booth, A. N., Robbins, D. J., Ribelin, W. E., and DeEds, F.,** Effect of raw soybean meal and amino acids on pancreatic hypertrophy in rats, *Proc. Soc. Exp. Biol. Med.,* 104, 681, 1960.
23. **Rackis, J. J.,** Physiological properties of soybean trypsin inhibitors and their relationship to pancreatic hypertrophy and growth inhibition of rats, *Fed. Proc.,* 24, 1488, 1965.
24. **Gertler, A., Birk, Y., and Bondi, A.,** A comparative study of the nutritional and physiological significance of pure soybean trypsin inhibitors and of ethanol-extracted soybean meals in chicks and rats, *J. Nutr.,* 91, 358, 1967.
25. **Crass, R. A. and Morgan, R. G. H.,** The effect of long term feeding of soya-bean flour diets on pancreatic growth in the rat, *Br. J. Nutr.,* 47, 119, 1982.

26. **McGuinness, E. E., Morgan, R. G. H., Levison, D. A., Frape, D. L., Hopwood, D., and Wormsley, K. G.,** The effects of long-term feeding of soya flour on the rat pancreas, *Scand. J. Gastroenterol.*, 15, 497, 1980.
27. **McGuinness, E. E., Morgan, R. G. H., Levison, D. A., Hopwood, D., and Wormsley, K. G.,** Interaction of azaserine and raw soya flour on the rat pancreas, *Scand. J. Gastroenterol.*, 16, 49, 1981.
28. **McGuinness, E. E., Hopwood, D., and Wormsley, K. G.,** Further studies of the effects of raw soya flour on the rat pancreas, *Scand. J. Gastroenterol.*, 17, 273, 1982.
29. **Longnecker, D. S. and Curphey, T. J.,** Adenocarcinoma of the pancreas in azaserine-treated rats, *Cancer Res.*, 35, 2249, 1975.
30. **Morgan, R. G. H., Levinson, D. A., Hopwood, D., Saunders, J. H. B., and Wormsley, K. G.,** Potentiation of the action of azaserine on the rat pancreas by raw soya bean flour, *Cancer Lett.*, 3, 87, 1977.
31. **Pour, P., Krüger, F. W., Althoff, J., Cardesa, A., and Mohr, U.,** A new approach for induction of pancreatic neoplasms, *Cancer Res.*, 35, 2259, 1975.
32. **Pour, P., Salmasi, S., Runge, R., Gingell, R., Wallcave, L., Nagel, D., and Stepan K.,** Carcinogenicity of N-Nitrosobis(2-hydroxypropyl)amine and N-nitrosobis(2-oxopropyl)amine in the MRC rats, *J. Natl. Cancer Inst.*, 63, 181, 1979.
33. **Levison, D. A., Morgan, R. G. H., Brimacombe, J. S., Hopwood, D., Coghill, G., and Wormsley, K. G.,** Carcinogenic effects of di(2-hydroxypropyl)-nitrosamine (DHPN) in male Wistar rats: promotion of pancreatic cancer by a raw soya flour diet, *Scand. J. Gastroenterol.*, 14, 217, 1979.
34. **Crass, R. A. and Morgan, R. G. H.,** Rapid changes in pancreatic DNA, RNA and protein in the rat during pancreatic enlargement and involution, *Int. J. Vit. Nutr. Res.*, 51, 85, 1981.
35. **Crass, R. A.,** The Effects of Soya Flour Diets and Azaserine on the Growth and Function of the Rat Pancreas, Ph.D. thesis, University of Western Australia, 1983.
36. **Rackis, J. J., Smith, A. K., Nash, A. M., Robbins, D. J., and Booth, A. N.,** Feeding studies on soybeans. Growth and pancreatic hypertrophy in rats fed soybean meal fractions, *Cereal Chem.*, 40, 531, 1963.
37. **Lyman, R. L., Wilcox, S. S., and Monsen, E. R.,** Pancreatic enzyme secretion produced in the rat by trypsin inhibitors, *Am. J. Physiol.*, 202, 1077, 1962.
38. **Kwaan, H. C., Kok, P., and Astrup, T.,** Impairment of growth and pancreatic hypertrophy in rats fed trypsin inhibitor from raw peanuts, *Experentia*, 24, 1125, 1968.
39. **Snook, J. W.,** Factors in whole-egg protein influencing dietary induction of increases in enzyme and RNA levels in rat pancreas, *J. Nutr.*, 97, 286, 1969.
40. **Arnesjö, B., Ihse, I., Lundquist, I., and Qvist, I.,** Effects on exocrine and endocrine rat pancreatic functions of bovine lung trypsin inhibitor administered perorally, *Scand. J. Gastroenterol.*, 8, 545, 1973.
41. **Geratz, J. D. and Hurt, J. P.,** Regulation of pancreatic enzyme levels by trypsin inhibitors, *Am. J. Physiol.*, 219, 705, 1970.
42. **Green, G. M. and Lyman, R. L.,** Feedback regulation of pancreatic enzyme secretion as a mechanism for trypsin inhibitor-induced hypersecretion in rats, *Proc. Soc. Exp. Biol. Med.*, 140, 6, 1972.
43. **Green, G. M., Olds, B. A., Matthews, G., and Lyman, R. L.,** Protein, as a regulator of pancreatic enzyme secretion in the rat, *Proc. Soc. Exp. Biol. Med.*, 142, 1162, 1973.
44. **Schneeman, B. O. and Lyman, R. L.,** Factors involved in the intestinal feedback regulation of pancreatic enzyme secretion in the rat, *Proc. Soc. Exp. Biol. Med.*, 148, 897, 1975.
45. **Brand, S. J. and Morgan, R. G. H.,** The release of rat intestinal cholecystokinin after oral trypsin inhibitor measured by bio-assay, *J. Physiol.*, 319, 325, 1981.
46. **Adrian, T. E., Pasquali, C., Pescosta, F., Bacarese-Hamilton, A. J., and Bloom, S. R.,** Soya-induced pancreatic hypertrophy and rise of circulating cholecystokinin, *Gut*, 23, 889, 1982.
47. **Fölsch, U. R., Schafmayer, A., Becker, H. D., and Cruetzfeldt, W.,** Elevated plasma CCK-concentrations in exocrine pancreatic atrophy in the rat, *Dig. Dis.*, 28, 934, 1983.
48. **Denda, A., Inui, S., Sunagawa, M., Takahashi, S., and Konishi, Y.,** Enhancing effect of partial pancreatectomy and ethionine-induced pancreatic regeneration on the tumorigenesis of azaserine in rats, *Gann*, 69, 633, 1978.
49. **McGuinness, E. E., Hopwood, D., and Wormsley, K. G.,** Potentiation of pancreatic carcinogenesis in the rat by DL-ethionine-induced pancreatitis, *Scand. J. Gastroenterol.*, 18, 189, 1983.
50. **Pour, P. M., Donnelly, T., Stepan, K., and Muffly, K.,** Modification of pancreatic carcinogenesis in the hamster model. II. The effect of partial pancreatectomy, *Am. J. Pathol.*, 110, 75, 1983.
51. **Kondo, S.,** Carcinogenesis in relation to the stem-cell-mutation hypothesis, *Differentiation*, 24, 1, 1983.
52. **Cairns, J.,** Mutation selection and the natural history of cancer, *Nature (London)*, 255, 197, 1975.
53. **Rackis, J. J.,** Biological active components, in *Soybeans: Chemistry and Technology*, Vol. 1, Smith, A. K. and Circle, S. J., Eds., AVI Publishing, Westport, Conn., 1972, 158, Chap. 6.
54. **Liener, I. E. and Kakade, M. L.,** Protease inhibitors, in *Toxic Constituents of Plant Foodstuffs*, Liener, I. E., Ed., Academic Press, New York, 1969, 8.

55. **Johnson, L. R.,** Effects of gastrointestinal hormones on pancreatic growth, *Cancer,* 47, 1640, 1981.
56. **Dembinski, A., Gregory, H., Konturek, S. J., and Polanski, M.,** Trophic action of epidermal growth factor on the pancreas and gastroduodenal mucosa in rats, *J. Physiol.,* 325, 35, 1982.
57. **Krogdahl, A. and Holm, H.,** Pancreatic proteinases from man, trout, rat, pig, cow, chicken, mink and fox. Enzyme activities and inhibition by soyabean and lima bean proteinase inhibitors, *Comp. Biochem. Physiol.,* 74B, 403, 1983.
58. **Temler, R. S., Simon, E., and Amiguet, P.,** Comparison of the interactions of soya bean protease inhibitors with rat pancreatic enzymes and human trypsin, *Enzyme,* 30, 105, 1983.
59. **Patten, J. R., Richards, E. A., and Pope, H., II,** The effect of raw soybean on the pancreas of adult dogs, *Proc. Soc. Exp. Biol. Med.,* 137, 59, 1971.
60. **Sale, J. K., Goldberg, D. M., Fawcett, A. N., and Wormsley, K. G.,** Chronic and acute studies indicating absence of exocrine pancreatic feedback inhibition in dogs, *Digestion,* 15, 540, 1977.
61. **Hasdai, A. and Liener, I. E.,** Growth, digestibility and enzymatic activities in the pancreas and intestines of hamsters fed raw and heated soy flour, *J. Nutr.,* 113, 662, 1983.
62. **Ihse, I., Lilja, P., and Lundquist, I.,** Feedback regulation of pancreatic enzyme secretion by intestinal trypsin in man, *Digestion,* 15, 303, 1977.
63. **Jacobson, D., Tillman, R., Toskes, P., and Curington, C.,** Acute and chronic feedback inhibition of pancreatic exocrine secretion in human subjects, *Gastroenterology,* 82, 1091, 1982.
64. **Hotz, J., Ho, S. B., Go, V. L. W., and DiMagno, E. P.,** Short-term inhibition of duodenal tryptic activity does not affect human pancreatic, biliary, or gastric function, *J. Lab. Clin. Med.,* 101, 488, 1983.
65. **Dlugosz, J., Fölsch, U. R., and Creutzfeldt, W.,** Inhibition of intraduodenal trypsin does not stimulate exocrine pancreatic secretion in man, *Digestion,* 26, 197, 1983.
66. **Doell, B. H., Ebden, C. J., and Smith, C. A.,** Trypsin inhibitor activity of conventional foods which are part of the British diet and some soya products, *Qualitas Plantarum,* 31, 139, 1981.
67. **Carroll, K. K. and Khor, H. T.,** Dietary fat in relation to tumorigenesis, *Prog. Biochem. Pharmacol.,* 10, 308, 1975.
68. **Birt, D. F., Salmasi, S., and Pour, P. M.,** Enhancement of experimental pancreatic cancer in Syrian golden hamsters by dietary fat, *J. Natl. Cancer Inst.,* 67, 1327, 1981.
69. **Birt, D. F., Stepan, K. R., and Pour, P. M.,** Interaction of dietary fat and protein on pancreatic carcinogenesis in Syrian golden hamsters, *J. Natl. Cancer Inst.,* 71, 355, 1983.
70. **Roebuck, B. D. and Longnecker, D. S.,** Dietary lipid promotion of azaserine-induced pancreatic tumors in the rat, in *Diet and Cancer: From Basic Research to Policy Implications,* Roe, D., Ed., Alan R. Liss, New York, 1983.
71. **Boorman, G. and Eustis, S.,** Proliferative lesions of the exocrine pancreas in male F-344/N rats, *Environ. Health Perspectives,* 56, 213, 1984.
72. **Weisblum, B., Herman, L., and Fitzgerald, P. J.,** Changes in pancreatic acinar cells during protein deprivation, *J. Cell Biol.,* 12, 313, 1962.
73. **Sidransky, H.,** Chemical and cellular pathology of experimental acute amino acid deficiency, *Methods Achiev. Exp. Pathol.,* 6, 1, 1972.
74. **Sidransky, H.,** Chemical pathology of nutritional deficiency induced by certain plant proteins, *J. Nutr.,* 71, 387, 1960.
75. **Campbell, T. C., Hayes, J. R., Merrill, A. H., Jr., Maso, M., and Goetchius, M.,** The influence of dietary factors on drug metabolism in animals, *Drug Metab. Rev.,* 9, 173, 1979.
76. **Roebuck, B. D., Yager, J. D., Jr., and Longnecker, D. S.,** Dietary modulation of azaserine-induced pancreatic carcinogenesis in the rat, *Cancer Res.,* 41, 888, 1981.
77. **Pour, P. M., Birt, D. F., Salmasi, S. Z., and Götz, U.,** Modifying factors in pancreatic carcinogenesis in the hamster model. I. Effect of protein-free diet fed during the early stages of carcinogenesis, *J. Natl. Cancer Inst.,* 70, 141, 1983.
78. **Pour, P. M. and Birt, D. F.,** Modifying factors in pancreatic carcinogenesis in the hamster model. IV. Effects of dietary protein, *J. Natl. Cancer Inst.,* 71, 347, 1983.
79. **O'Conner, T. P., Youngman, L. D., and Campbell, T. C.,** Effect of selenium on development of L-azaserine induced preneoplastic abnormal acinar cell nodules in rat pancreas, *Fed. Proc.,* 42, 670, 1983.
80. **Ip, C.,** Prophylaxis of mammary neoplasia by selenium supplementation in the initiation and promotion phases of chemical carcinogenesis, *Cancer Res.,* 41, 4386, 1981.
81. **Lawson, T. and Birt, D. F.,** Enhancement of the repair of carcinogen-induced DNA damage in the hamster pancreas by dietary selenium, *Chem. Biol. Interact.,* 45, 95, 1983.
82. **Birt, D. F., Lawson, T. A., Julius, A. D., Runice, C. E., and Salmasi, S.,** Inhibition by dietary selenium of colon cancer induced in the rat by bis(2-oxopropyl)nitrosamine, *Cancer Res.,* 42, 4455, 1982.
83. **Combs, G. F. Jr. and Bunk, M. J.,** The role of selenium in pancreatic function, in *Selenium in Biology and Medicine,* Spallholz, J. E., Martin, J. L., and Ganther, H. E., Eds., AVI Publishing, Westport, Conn., 1981, 70.

84. **Raica, N., Jr., Stedham, M. A., Herman, Y. F., and Sauberlich, H. E.,** Vitamin A deficiency in germ-free rats, in *The Fat-Soluble Vitamins,* DeLuca, H. F., and Suttie, J. W., Eds., University of Wisconsin Press, Madison, 1969, 283.
85. **Longnecker, D. S., Curphey, T. J., Kuhlmann, E. T., and Roebuck, B. D.,** Inhibition of pancreatic carcinogenesis by retinoids in azaserine-treated rats, *Cancer Res.,* 42, 19, 1982.
86. **Longnecker, D. S., Kuhlmann, E. T., and Curphey, T. J.,** Divergent effects of retinoids on pancreatic and liver carcinogenesis in azaserine-treated rats, *Cancer Res.,* 43, 3219, 1983.
87. **Longnecker, D. S., Kuhlmann, E. T., and Curphey, T. J.,** Effects of four retinoids in N-nitrosobis(2-oxopropyl)amine-treated hamsters, *Cancer Res.,* 43, 3226, 1983.
88. **Birt, D. F., Davies, M. H., Pour, P. M., and Salmasi, S.,** Lack of inhibition by retinoids of bis(2-oxopropyl)nitrosamine-induced carcinogenesis in Syrian hamsters, *Carcinogenesis,* 4, 1216, 1983.
89. **Birt, D. F., Sayed, S., Davies, M. H., and Pour, P.,** Sex differences in the effects of retinoids on carcinogenesis by N-nitrosobis(2-oxopropyl)amine in Syrian hamsters, *Cancer Lett.,* 14, 13, 1981.
90. **Fölsch, U. R. and Creutzfeldt, W.,** Pancreatic duct cells in rats: secretory studies in response to secretin, cholecystokinin-pancreozymin and gastrin in vivo, *Gastroenterology,* 73, 1053, 1977.
91. **Lombardi, B.,** Influence of dietary factors on the pancreatotoxicity of ethionine, *Am. J. Pathol.,* 84, 633, 1976.
92. **Shinozuka, H., Katyal, S. L., and Lombardi, B.,** Azaserine carcinogenesis: organ susceptibility change in rats fed a diet devoid of choline, *Int. J. Cancer,* 22, 36, 1978.
93. **Zurlo, J., Roebuck, B. D., Rutkowski, J. V., Curpey, T. J., and Longnecker, D. S.,** Effect of pyridoxal deficiency on pancreatic DNA damage and nodule induction by azaserine, *Carcinogenesis,* 5, 555, 1984.
94. **Tweedie, J. H., Reber, H. A., Pour, P. M., and Pounder, D. M.,** Protective effect of ethanol on the development of pancreatic cancer, *Surg. Forum,* 32, 222, 1981.
95. **Pour, P. M., Reber, H. A., and Stepan, K.,** Modification of pancreatic carcinogenesis in the hamster model. XII. Dose-related effect of ethanol, *J. Natl. Cancer Inst.,* 71, 1085, 1983.
96. **Rackis, J.,** personal communication.
97. **Wormsley, K. G.,** personal communication.
98. **Longnecker, D. S., Curphey, T. J., Kuhlmann, E. T., Roebuck, B. D., and Neff, R. K.,** Effects of retinoids in N-nitrosobis (α-oxopropyl) amine-treated hamsters, in preparation.

Chapter 3

THE EPIDEMIOLOGY OF LARGE BOWEL CANCER

Ole Møller Jensen

TABLE OF CONTENTS

I. INTRODUCTION

Large bowel cancer is the common denominator for cancer of the colon and rectum, which by the World Health Organization Disease Classification[1] are regarded as two separate entities, ICD-153 and ICD-154. These are diseases of the affluent, westernized cultures of Europe, North America, and Australia. In most countries of these parts of the world colorectal cancer ranks among the five most important cancers,[2] and in these high-risk cultures some 5% of the population will develop the diseases before they reach the age of 75.[2] Approximately 3% of all deaths are caused by colorectal cancer in Europe and North America.

Colorectal cancer thus represents a major public health problem which has attracted attention both from cancer epidemiology and experimental cancer research. In spite of this most of our present understanding of the underlying determinants of the disease still rests on observations of patterns of disease occurrence in man. From the variation in colorectal cancer incidence in human populations we attempt to deduce biologically plausible hypotheses concerning etiologies, which can be further examined by epidemiological studies in man often drawing extensively on the collaboration between the epidemiologist and laboratory researchers in the conduct of biochemical epidemiology investigations.

This chapter deals with the occurrence of colorectal cancer in human populations.

II. VARIATION WITH AGE AND SEX

Like most neoplasms of an epithelial origin cancers of the various parts of the large bowel increase exponentially with age as illustrated in Figure 1. This relationship with age, i.e., the slope of the age incidence curve for cancer of the colon and rectum is quite similar for a number of counties, whether high- or low-risk areas.[3] The relationship with age is likely to result from the continuous action of a constantly present carcinogenic stimulus on the large bowel.[4] Hypotheses concerning the nature of the carcinogenic influences should be in line with this observation.

Figure 2 illustrates that the incidence in both sexes decreases from the ascending colon to the transverse and the descending colon followed by a sharp increase in the sigmoid and rectum, Figure 1. The same general pattern emerges in both high- and low-risk areas,[5,6] although the sigmoid-rectum ratio is higher in North American than in European, South American, and Asian populations. Differences in the definition of the recto-sigmoid junction in various countries must be considered before drawing firm conclusions on this point.[6]

Figure 2 also shows that the age and sex relationship with incidence vary for the different anatomic parts of the colon. The figures are based on data from the Danish Cancer Registry for the period 1978 to 1980, where 9.6% of all male and 11.6% of all female colon cancers were not classified as to subsite. For all subsites the risk increases with increasing age, but for the right colon the female rates are higher (ascending) or at the same level (transverse) as male rates in older age groups. For the left colon (descending and sigmoid) male incidence in older ages lies above that of females, a phenomenon which becomes further visible for the rectum, Figure 1. The observed female preponderance of colon cancer in premenopausal age groups[7] is seen for cancer of the descending and sigmoid colon only.

Summarizing the incidence curves in Figure 1 as age-standardized rates a male-female ratio emerges for colon cancer of about 1.0, whereas it is about 1.5 for rectum cancer. These features of the sex ratio are found constantly around the world as shown in Table 1. The higher female than male rates at younger ages have been related to the occurrence of menopause, and it has been suggested that female sex hormones may play a role in colon cancer etiology by influencing bile secretion.[7] The different sex ratio for cancer of the colon and cancer of the rectum clearly suggest that etiological factors for malignant neoplasms of these two parts of the large bowel may at least in part be different or carry different weights.

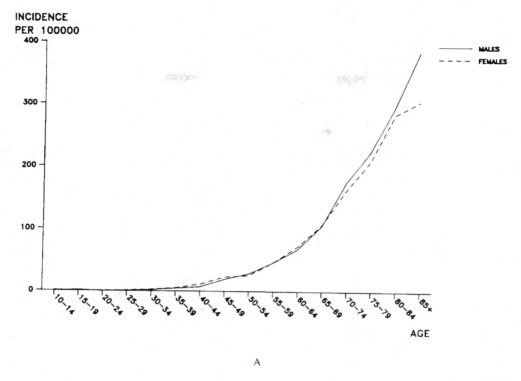

A

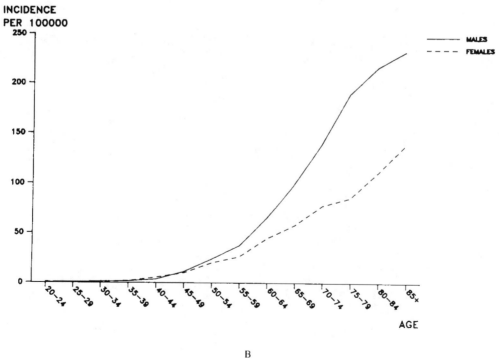

B

FIGURE 1. Age incidence curve for cancer of the colon (A) and cancer of the rectum (B). Denmark 1978 to 1980.

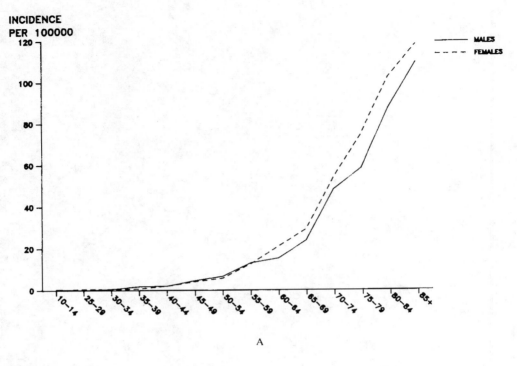

A

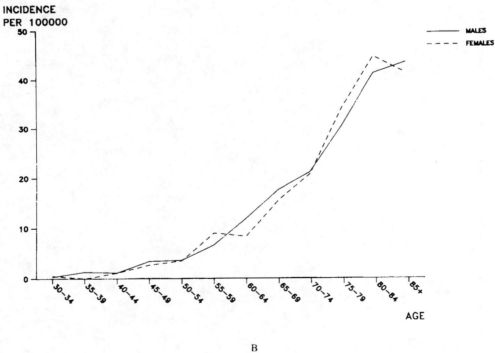

B

FIGURE 2.   Age incidence curve for cancer of subsites of the colon, Denmark, 1978 to 1980. (A) Ascending;
(B) transverse colon including flexures; (C) descending colon; (D) sigmoid colon including recto sigmoid.

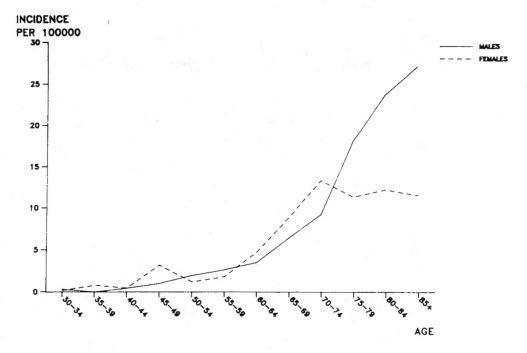

FIGURE 2C

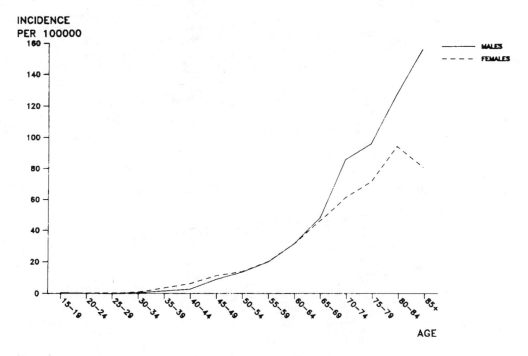

FIGURE 2D

Table 1
AGE-STANDARDIZED (WORLD POPULATION) INCIDENCE
RATES PER 100,000 AND MALE-FEMALE RATIOS OF
COLON AND RECTUM CANCER FOR SELECTED REGIONS
OF THE WORLD

Region	Colon (ICD-153)			Rectum (ICD-154)		
	Male	Female	M/F	Male	Female	M/F
Africa						
Senegal, Dakar	1.3	1.3	1.0	3.0	2.9	1.0
Americas						
Canada, Manitoba	32.9	31.5	1.0	19.1	14.5	1.3
Canada, Quebec	25.9	29.0	0.9	18.6	13.6	1.4
Colombia, Cali	5.4	6.8	0.8	5.1	3.5	1.5
U.S., Connecticut	37.8	33.8	1.1	24.7	15.7	1.6
U.S., New York	36.5	34.2	1.1	22.4	15.6	1.4
U.S., Utah	26.5	27.0	1.0	14.5	10.5	1.4
Asia						
India, Bombay	4.8	4.4	1.1	4.7	4.0	1.2
Israel, all Jews	15.7	16.7	0.9	17.9	19.3	0.9
Japan, Miyagi	11.0	10.6	1.0	11.7	9.8	1.2
Singapore, Chinese	21.3	20.1	1.1	20.0	11.3	1.8
Europe						
Denmark	21.7	24.6	0.9	22.1	14.8	1.5
Finland	10.9	12.0	0.9	11.1	8.9	1.2
Germany, D.R.	15.3	15.8	1.0	18.6	15.6	1.2
Norway	18.4	20.8	0.9	15.2	12.6	1.2
Poland, Warsaw City	16.8	12.7	1.3	13.2	9.8	1.3
Poland, Warsaw Rural	9.1	6.6	1.4	8.7	5.3	1.6
Spain, Zaragoza	8.3	8.2	1.0	8.2	8.9	0.9
Sweden	19.2	20.4	0.9	13.3	10.6	1.3
Switzerland, Geneva	26.4	20.4	1.3	17.0	14.1	1.2
U.K., Birmingham	20.7	22.7	0.9	22.5	14.4	1.6
Yugoslavia, Slovenia	12.7	9.7	1.3	16.0	12.3	1.3
Australia						
Australia, NSW	29.3	26.1	1.1	19.3	13.5	1.4
New Zealand, Non-Maori	37.9	43.9	0.9	25.6	19.1	1.3

From Waterhouse, J., Muir, C. S., Shanmugaratnam, K., and Powell, J., Cancer Incidence in Five Continents, Vol. 4, IARC Scientific Publ. No. 42, International Agency for Research on Cancer, Lyon, 1982.

III. VARIATION WITHIN COUNTRIES

Geographical variations in incidence of large bowel cancer are known to exist within countries where mortality or incidence has been recorded and described for small areas.[8-10] Such variations are typically much less than the variation observed between countries. In general, higher rates are seen in more affluent parts, and within a given country urban rates tend to be slightly higher than rates for rural areas, as illustrated with data from Denmark,[11] Finland,[12] and Norway[13] in Table 2.

In high incidence countries of North America and western Europe no marked differences in the incidence of colon cancer are associated with socioeconomic status, Table 3.[14] Some studies point however towards sigmoid cancer occurring slightly more often in higher socioeconomic classes than in lower ones, whereas the opposite trend is observed for cancer of the rectum.[14] In low incidence areas like Hong Kong[15] and most of South America,[16] studies show a clear trend with socioeconomic class for the risk of colon cancer (Table 3).

Table 2
URBAN-RURAL RATE — RATIOS
FOR COLON AND RECTUM CANCER
IN SCANDINAVIA

	Colon		Rectum	
Country	Male	Female	Male	Female
Denmark[a]	1.4	1.2	1.3	1.3
Finland[b]	2.1	1.6	1.4	1.3
Norway[c]	1.4	1.2	1.2	1.5

[a]   Copenhagen/rural.
[b]   Biggest towns/rural.
[c]   Urban/rural.

Table 3
SOCIAL CLASS DIFFERENCES
IN COLON CANCER IN HIGH
(U.K.) AND LOW (HONG KONG)
INCIDENCE AREAS — MALES

England and Wales		Hong Kong	
Social class	SMR[a]	Relative risk	Income level
I	104		
II	100	1.9	High
III N	106		
III M	106	1.5	Medium
IV	100		
V	110	1.0	Low

[a]   Smr, Standardized mortality ratio.

Incidence differences are also recorded both for colon and for rectum cancer in various ethnic groups living in close proximity in the same geographic locality.[2] There is thus a two- to three-fold variation between Chinese, Malays, and Indians living in Singapore[2] and a three-fold difference between persons with and without Spanish surnames living in the southern U.S.,[2] Table 4. In New Zealand rates among non-Maories are three to four times higher than in the Maories.[2]

The study of colorectal cancer in religious groups has shown a markedly lower risk of 60 to 70% among Seventh Day Adventists in California compared with other white Americans.[17] The Adventists are to a large degree lacto-ovo vegetarians abstaining also from tobacco and alcohol. There has been some concern that the decreased risk could be explained by selection or by confounding, but there is no indication that better medical care than provided to the average American is the explanation. The investigation of cancer incidence among male Seventh Day Adventists in Denmark corroborates the findings from the U.S. with regard to colon cancer where the risk is only 10% ($p < 0.05$) of that recorded in the general population, whereas rectum cancer is similar to that of the Danish population at large.[18] As the decreased risk of colon cancer has been ascribed to the Seventh Day Adventist's habit of abstaining from eating meat it is somewhat puzzling that in the U.S. a

Table 4
ETHNIC GROUP DIFFERENCES IN COLON AND RECTUM CANCER INCIDENCE (AGE-STANDARDIZED WORLD POPULATION) IN SELECTED AREAS

Region/population	Colon		Rectum	
	Males	**Females**	**Males**	**Females**
U.S., New Mexico				
American Indian	9.4	9.8	8.6	2.6
Spanish	18.3	15.3	8.4	9.4
Other white	33.9	27.8	15.8	10.6
Singapore				
Indian	6.7	9.8	9.5	12.1
Malay	8.7	5.2	9.0	7.8
Chinese	21.3	20.1	20.0	11.3
New Zealand				
Maori	11.4	13.7	14.8	11.9
Non-Maori	37.9	43.9	25.6	19.1

similar decrease in colorectal cancer risk is observed among Mormons.[19] It has been pointed out however, that the healthy lifestyle of the Mormons includes the eating of whole-grain bread and large amounts of cereals, which may provide a high daily intake of dietary fiber suggested to lower the risk of colon cancer.[20-23]

Occupational exposure in general has not been associated with increased risk of colon cancer, except for a number of studies which point to a two- to threefold increase in colon cancer risk in relation to asbestos exposure.[24]

IV. INTERNATIONAL VARIATION

Cancer of the large bowel is primarily a disease of the industrially developed parts of the world. The highest rates of colon cancer in both males and females are observed in North America and in western Europe, and much lower rates are generally seen in eastern Europe, Asia, Africa, and South America, Figure 3. The low rates in Japan point to an international association with the "western lifestyle" rather than industrialization as such.

Although some of the international variations may be explained by diagnostic differences and the availability and use of medical facilities, it is unlikely that this would account for all of the differences seen in the truncated rates which describe the variation among persons 35 to 64 years of age. Within Scandinavia where good cancer incidence data exist for more than 30 years a two- to threefold incidence difference is observed between Denmark and Finland;[25] an evaluation of this difference proved it to be valid.[26]

The incidence of cancer of the rectum varies somewhat less internationally than cancer of the colon, Figure 4, but there is a positive correlation between the incidence of colon and rectum cancer, Figure 5. Thus in general when colon cancer rates are high so are the rates for cancer of the rectum.

V. MIGRANT POPULATIONS

The study of large bowel cancer in migrant populations shows that international differences cannot be ascribed to variation in genetic background, but must be ascribed to environmental influences.[27] Table 5 shows the shift in risk of both colon and rectum cancer in Japanese

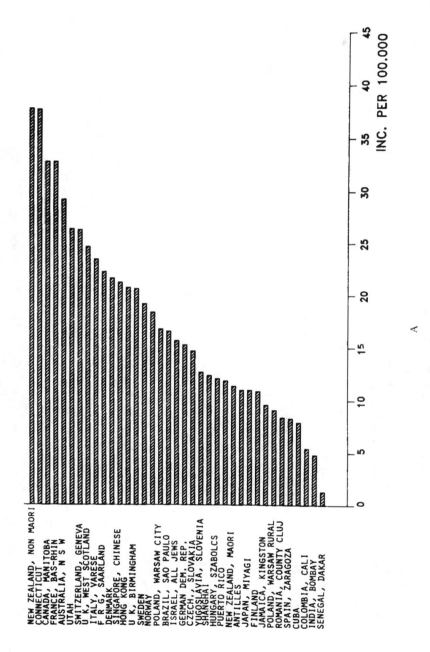

FIGURE 3.   International variation in age-standardized, truncated (ages 35 to 64) incidence of colon cancer in males (A) and females (B).

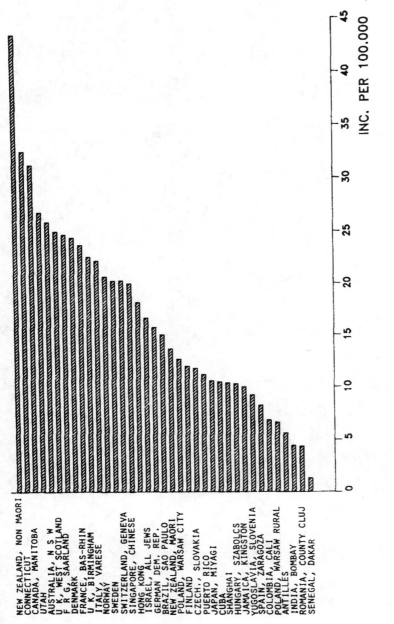

FIGURE 3B

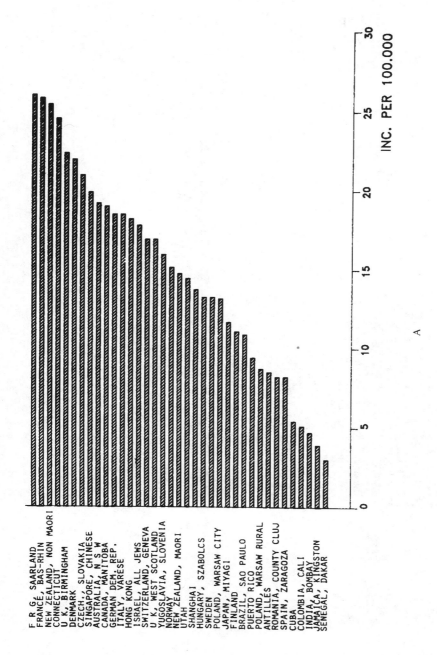

FIGURE 4. International variation in age-standardized, truncated (ages 35 to 64) incidence of rectum cancer in males (A) and females (B).

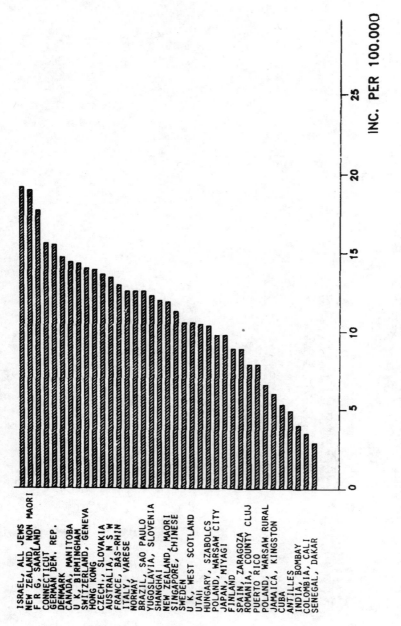

FIGURE 4B

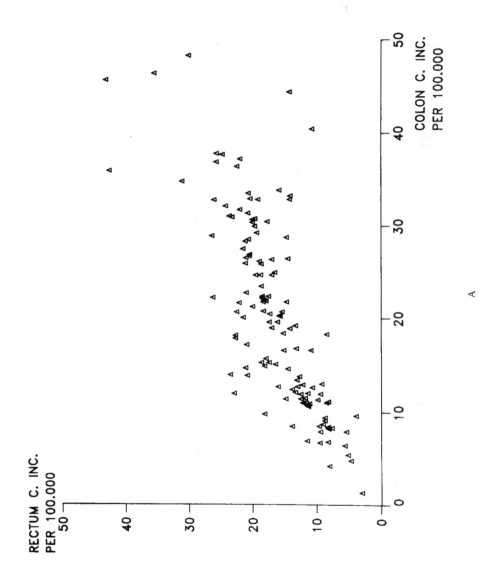

FIGURE 5. Relationship between rectum cancer and colon cancer incidence rates (standardized, truncated to ages 35 to 64) for male (A) and female (B) populations. (From Waterhouse, J., Muir, C. S., Shanmugaratnam, K., and Powell, J., Cancer Incidence in Five Continents, Vol. 4, IARC Scientific Publ. No. 42, International Agency for Research on Cancer, Lyon, 1982.)

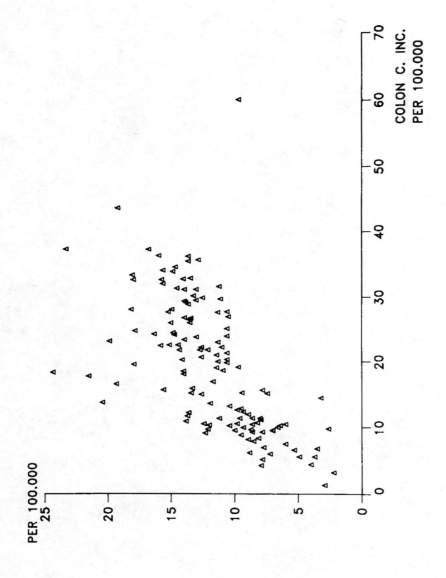

FIGURE 5B

Table 5
AGE-STANDARDIZED (WORLD POPULATION)
INCIDENCE PER 100,000 OF COLON AND RECTUM
CANCER IN IMIGRANTS

Population group/region	Colon		Rectum	
	Male	Female	Male	Female
Japanese				
Japan, Miyagi	11.0	10.6	11.7	9.8
Hawaii, Japanese	34.9	30.3	31.0	13.2
Blacks				
Senegal, Dakar	1.3	1.3	3.0	2.9
U.S., Bay Area, Black	44.5	33.5	14.2	18.0
Israel Jews				
Born Africa, Asia	14.9	17.9	18.1	21.5
Born European, American	8.5	8.4	9.6	8.1

who move from low-risk Japan to high-risk U.S. (Figures 3 and 4). Table 5 also indicates that Israeli Jews retain the rates of their countries of origin, as the Israel Cancer Registry records high rates among Jews born in Europe and U.S. and low rates among those born in Africa and Asia.[2] Much higher incidence rates are recorded among U.S. Blacks than among African Blacks.[2]

Studies of the cancer risk in various generations of Japanese migrants to the U.S.[27] show that the rates approach the high U.S. incidence for colon cancer within the first generation. A convergence of colon cancer incidence rates to those of Australia has been demonstrated for European migrants to that country.[28] Irrespective of whether they arrive from high-risk Scotland or from low-risk Poland, Yugoslavia, Greece, or Italy, the rates approach those of Australia the longer these migrants have lived in the country, probably reflecting the effect of acculturation.

VI. TIME TRENDS

The study of changes in cancer risk over time requires long-standing, high quality cancer registration for the reliable acquisition of incidence data. Although such data may be subject to changing clinical definitions of disease,[29] they are uninfluenced by changes in survival and in habits related to death certification.[30,31] The latter is often subject to errors, which may effect the mortality statistics.[32] Tables 6 and 7 show the development in age-standardized incidence rates for cancer of the colon and rectum in selected populations included in Volumes 1 to 4 of Cancer Incidence in Five Continents.[2,33-35] Both sites are characterized by a slight increase over the approximately 15 years represented. For colon cancer an annual average increase of some 1.5 to 3% is seen irrespective of the level of the overall incidence rate. Japan (Miyagi) is a notable exception, as the incidence in males has gone up from 3.9 to 8.3 per 100,000 and in females from 4.1 to 7.3 per 100,000 (Table 6). For rectum cancer a similar overall upward trend is apparent. For this site it seems that rates are increasing faster in low than in high incidence areas (above 10% per 100,000), thus tending to minimize international differences as time passes.

VII. CONCLUSIONS

The study of the international variation of colon and rectum cancer in defined populations points to factors associated with a western lifestyle as important determinants of risk. Genetic factors as indicators of population risks are minimal as indicated by studies of migrants.

Table 6
TIME TRENDS IN AGE-STANDARDIZED INCIDENCE
OF COLON CANCER (WORLD POPULATION) IN
VARIOUS PARTS OF THE WORLD[2,33,34,35]

Region	Sex	I	II	III	IV	% Annual change
Canada, Alberta	M	15.2	14.0	17.1	18.1	+ 1.4
	F	16.1	18.8	18.5	18.0	+ 0.8
Canada, Manitoba	M	20.7	20.7	20.7	24.7	+ 1.4
	F	23.0	22.6	24.1	23.1	+ 0.3
Columbia, Cali	M	5.4	3.6	3.2	4.5	− 1.5
	F	3.0	4.0	3.4	5.4	+ 7.3
Jamaica, Kingston	M	8.5	10.6	9.1	8.7	+ 0.2
	F	9.4	8.6	7.9	7.5	− 1.3
U.S., Connecticut	M	25.4	26.7	30.1	32.3	+ 1.9
	F	25.4	26.7	26.1	26.4	+ 0.3
U.S., New York State	M	21.2	—	24.6	31.4	+ 3.2
	F	20.7	—	22.6	26.3	+ 1.8
Japan, Miyagi	M	3.9	4.1	5.6	8.3	+ 7.5
	F	4.1	4.0	5.4	7.3	+ 5.2
Denmark	M	15.5	16.2	17.5	19.0	+ 1.5
	F	17.9	17.5	18.7	18.7	+ 0.3
Finland	M	6.1	6.8	7.9	8.3	+ 2.6
	F	7.2	7.3	8.0	9.4	+ 2.2
Germany,	M	11.6	12.3	13.6	16.5	+ 3.0
F.R. Hamburg,	F	12.2	12.7	13.6	14.9	+ 1.6
Norway	M	10.5	12.0	12.7	14.3	+ 2.4
	F	10.2	11.6	13.0	14.5	+ 2.8
Sweden	M	13.4	14.7	15.8	16.3	+ 1.7
	F	13.0	14.0	14.7	15.1	+ 1.2
U.K., Birmingham	M	14.9	15.3	16.5	16.3	+ 0.7
	F	15.2	14.9	15.0	15.8	+ 0.3
Yugoslavia, Slovenia	M	5.0	5.7	6.0	7.7	+ 3.2
	F	4.6	5.3	6.0	6.6	+ 2.6
Hawaii						
White	M	18.9	19.3	23.9	25.3	+ 2.6
	F	24.0	27.5	22.9	17.8	− 2.0
Japanese	M	19.0	20.7	22.4	27.5	+ 3.4
	F	11.9	15.3	18.8	18.8	+ 4.5
Hawaiian	M	16.9	20.2	14.1	13.7	− 1.5
	F	12.9	12.1	16.9	14.0	+ 0.7

[a] Covering the periods of approximately: I: 1960—1962; II: 1964—1966; III: 1968—1972; IV: 1973—1977.

The study of twins recently corroborated this conclusion.[36] Such factors act on the bowel mucosa in such a way as to be associated with a constant increase in risk with progressing age. Possible sex differences in the age-specific rates for various parts of the large bowel must be considered when determining etiologies.

All of the recorded observations of variation in colorectal cancer incidence and mortality within countries indicate that the ambient environment is unimportant as a determinant of these cancers. Geographical differences are thus less important than the differences in incidence that may be seen in various groups of the population living in close proximity with each other. The socioeconomic differences in low-risk areas and the differences between ethnic and religious groups are likely indicators of differences in lifestyle associated with typical dietary patterns characterized by a highly refined diet.

Table 7
TIME TRENDS IN AGE-STANDARDIZED INCIDENCE OF RECTUM CANCER (WORLD POPULATION) IN VARIOUS PARTS OF THE WORLD[2,33,34,35]

Region	Sex	Time periods[a]				% Annual change
		I	II	III	IV	
Canada, Alberta	M	9.0	11.3	10.6	12.6	+ 2.9
	F	5.7	8.4	6.9	8.4	+ 3.4
Canada, Manitoba	M	11.1	11.7	13.7	14.1	+ 1.9
	F	9.5	8.3	9.0	9.6	+ 0.1
Colombia, Cali	M	2.2	3.9	3.1	3.4	+ 5.0
	F	1.9	3.0	3.3	2.3	+ 1.9
Jamaica, Kingston	M	4.3	5.8	4.9	3.7	− 0.9
	F	5.0	6.7	4.0	3.8	− 1.6
U.S., Connecticut	M	15.1	16.3	18.2	17.7	+ 1.2
	F	10.1	10.7	11.1	11.1	+ 0.7
U.S., New York State	M	13.4	—	13.7	16.8	+ 1.7
	F	8.6	—	8.7	10.4	+ 1.4
Japan, Miyagi	M	4.9	4.8	6.8	9.2	+ 5.9
	F	4.4	5.0	5.0	6.5	+ 3.2
Denmark	M	16.3	17.2	16.7	17.0	+ 0.3
	F	10.2	10.5	10.8	10.5	+ 0.2
Finland	M	6.3	7.6	7.7	8.7	+ 2.7
	F	5.1	6.4	6.1	6.6	+ 2.1
Germany,	M	10.4	11.0	12.0	12.9	+ 1.7
F. R., Hamburg	F	7.2	8.7	9.3	9.2	+ 2.0
Norway	M	6.5	6.8	10.1	11.7	+ 5.3
	F	4.2	4.7	7.1	8.0	+ 6.0
Sweden	M	10.3	10.4	10.5	10.9	+ 0.4
	F	6.2	6.9	6.9	7.2	+ 1.2
U.K., Birmingham	M	15.6	15.8	16.1	16.7	+ 0.7
	F	8.7	9.3	8.7	9.1	+ 0.3
Yugoslavia, Slovenia	M	6.5	8.2	11.4	11.2	+ 4.3
	F	4.8	6.2	6.6	7.9	+ 3.8
Hawaii						
White	M	9.0	12.6	13.5	14.1	+ 4.4
	F	8.5	9.8	12.0	8.9	+ 0.4
Japanese	M	11.4	11.7	16.3	21.4	+ 6.7
	F	7.3	9.3	10.1	8.8	+ 1.6′
Hawaiian	M	4.4	6.8	9.4	11.2	+11.9
	F	2.7	6.9	2.9	9.5	+19.4

[a] Covering the periods of approximately: I: 1960—1962; II: 1964—1966; III: 1968—1972; IV: 1973—1977.

In 1967 Wynder and Shigematsu[37] suggested that differences in fat intake may be responsible for the international distribution of colon cancer. From then on there has been a rapid development of etiologic hypotheses related to various aspects of the diet, as reported in the following chapter of this volume.

ACKNOWLEDGMENTS

Ms. Helene Hartmann Petersen and Mrs. Aase Falck kindly assisted with the preparation of the manuscript. This work was carried out with the financial support from the Danish Cancer Society.

REFERENCES

1. World Health Organization, Manual of the International Statistical Classification of Diseases, Injuries and Causes of Death, 9th revision, World Health Organization, Geneva, 1977.
2. **Waterhouse, J., Muir, C. S., Shanmugaratnam, K., and Powell, J.,** Incidence in Five Continents, Vol. 4, IARC Scientific Publ. No. 42, International Agency for Research on Cancer, Lyon, 1982.
3. **Cook, P. J., Doll, R., and Fellingham, A.,** A mathematical model for the age distribution of cancer in man, *Int. J. Cancer*, 4, 93, 1969.
4. **Muir, C. S. and Peron, Y.,** Special demographic situations, *Semin. Oncol.*, 3, 35, 1976.
5. **de Johg, U., Day, N. E., Muir, C. S., Barcley, T. H. C., Bras, G., Foster, F. H., Jussawalla, D. J., Kurihara, M., Linden, G., Martinez, I., Payne, P. M., Pedersen, E., Ringertz, N., and Shanmugaratnam, T.,** The distribution of cancer within the large bowel, *Int. J. Cancer*, 10, 463, 1972.
6. **Jensen, O. M.,** Colon cancer epidemiology, in *Experimental Colon Carcinogenesis*, Autrup, H. and Williams, G. M., Eds., CRC Press, Boca Raton, Fla., 1983.
7. **McMichael, A. J. and Potter, J. D.,** Reproduction, endogenous and exogenous sex hormones and colon cancer: a review and hypothesis, *J. Natl. Cancer Inst.*, 65, 1201, 1980.
8. **Mason, T. J., McKay, F. W., Hoover, R., Blot, W. J., and Fraumeni, J. F.,** Atlas of Cancer Mortality for U.S. Countries: 1950—1969, Publ. No. (NIH)75-780, U.S. Department of Health, Education and Welfare, Washington, D.C., 1975.
9. **Frentzel-Beyme, R., Leutner, R., Wagner, G., and Wiebelt, H.,** *Cancer Atlas of the Federal Republic of Germany*, Springer-Verlag, Basel, 1979.
10. **Gardner, M. J., Winther, P. D., Taylor, C. P., and Acheson, E. D.,** *Atlas of Cancer Mortality in England and Wales, 1968—1978*, John Wiley & Sons, Chichester, 1983.
11. Danish Cancer Registry, Cancer Incidence in Denmark 1973—1977, Danish Cancer Registry, Copenhagen, 1982.
12. **Teppo, L., Pukkala, E., Hakama, M., Hakkulinen, T., Herva, A., and Saxen, E.,** Way of life and cancer incidence in Finland, *Scand. J. Soc. Med.*, Suppl. 19, 1980.
13. Cancer Registry of Norway, in Cancer Incidence in Five Continents, Vol. 4, IARC Scientific Publ. No. 42, Waterhouse, J., Muir, C., Shanmugaratnam, K., and Powell, J., Eds., International Agency for Research on Cancer, Lyon, 1982.
14. Office of Population Censuses and Surveys, Occupational Mortality, The Registrar General Decennial Supplement for England and Wales, 1970 to 1972, Her Majesty's Stationary Office, London, 1978.
15. **Crowther, J. S., Drasar, B. S., Hill, M. J., MacLennan, R., Magnin, D., Peach, S., and Tech-Chan, C. H.,** Faecal steroids and bacteria in large bowel cancer in Hong Kong by socioeconomic groups, *Br. J. Cancer*, 34, 191, 1976.
16. **Haenszel, W., Correa, P., and Cuello, C.,** Social class differences among patients with large bowel cancer in Cali, Colombia, *J. Natl. Cancer Inst.*, 54, 1031, 1975.
17. **Phillips, R. L., Kuzma, J. W., and Lodz, T. M.,** Cancer mortality among comparable members versus non-members of the Seventh Day Adventist Church, in *Cancer Incidence in Defined Populations*, Banbury Report No. 4, Cairns, J., Lyon, J. L., and Skolnick, N., Eds., Cold Spring Harbor Laboratory, Cold Spring Harbor, N.Y., 1980, 93.
18. **Jensen, O. M.,** Cancer risk among Danish male Seventh Day Adventists and other temperance society members, *J. Natl. Cancer Inst.*, 70, 1011, 1983.
19. **Lyon, J. L., Gardner, J. W., and West, D. W.,** Cancer risk and life-style: cancer among Mormons from 1967—1975, in *Cancer Incidence in Defined Populations*, Banbury Report No. 4, Cairns, J., Lyon, J. L., and Skolnick, N., Eds., Cold Spring Harbor Laboratory, Cold Spring Harbor, N.Y., 1980, 3.
20. IARC Microecology Group, Dietary fibre, transit time, faecal bacteria, steroids and colon cancer in 2 Scandinavian populations, *Lancet*, 2, 207, 1977.
21. **Jensen, O. M., MacLennan, R., Wahrendorff, J., and IARC,** Large Bowel Group, Large bowel cancer in Scandinavia in relation to diet and faecal characteristics, *Nutr. Cancer*, 4, 1982.
22. **Modan, B., Marell, V., Lubin, F., Modan, M., Greenberg, R. A., and Graham, S.,** Low fibre intake as an aetiologic factor in cancer of the colon, *J. Natl. Cancer Inst.*, 55, 15, 1975.
23. **Reddy, B. S., Hedges, A. R., Laakso, K., and Wynder, E. L.,** Metabolic epidemiology of large bowel cancer: faecal bulk and constituents of high risk North American and low risk Finnish population, *Cancer*, 42, 2831, 1978.
24. IARC, Monograph on the Evaluation of the Carcinogenic Risk of Chemicals to Humans. Asbestos, International Agency for Research on Cancer, Lyon, 1977.
25. **Jensen, O. M.,** Cancer of the large bowel, a 3-fold variation between Denmark and Finland, *Nutr. Cancer*, 4, 20, 1982.
26. **Jensen, O. M., Mosbech, J., Salaspuro, M., and Ihamaki, T.,** A comparative study of the diagnostic basis for cancer of the colon and cancer of the rectum in Denmark and Finland, *Int. J. Epidemiol.*, 3, 183, 1974.

27. **Haenszel, W.,** Cancer mortality among the foreign born in the United States, *J. Natl. Cancer Inst.,* 26, 37, 1961.

28. **McMichael, A. J., McCall, M. G., Hartshorne, J. M., and Woodlings, T. L.,** Patterns of gastrointestinal cancer in European migrants to Australia: the role of dietary change, *Int. J. Cancer,* 25, 431, 1980.

29. **Doll, R. and Peto, R.,** The causes of cancer: quantitative estimates of avoidable risks in the USA today, *J. Natl. Cancer Inst.,* 66, 1193, 1981.

30. **Clemmesen, J.,** Statistical studies in the aetiology of malignant neoplasma. I, *Acta Pathol. Microbiol. Scand.,* Suppl. 174, 1964.

31. **Devesa, S., Pollack, E. S., and Young, J. L.,** Assessing the validity of observed cancer incidence trends, *Am. J. Epidemiol.,* 119, 274, 1984.

32. **Percy, C., Stanek, E., and Gloeckler, L.,** Accuracy of cancer death certificates and its effect on cancer mortality statistics, *Am. J. Public Health,* 71, 242, 1981.

33. **Doll, R., Payne, P., and Waterhouse, J.,** *Cancer Incidence in Five Continents,* Vol. 1, International Union against Cancer, Springer-Verlag, Berlin, 1966.

34. **Doll, R., Muir, C. S., and Waterhouse, J.,** *Cancer Incidence in Five Continents,* Vol. 2, International Union against Cancer, Springer-Verlag, Berlin, 1970.

35. **Waterhouse, J., Muir, C. S., Correa, P., and Powell, J.,** Cancer Incidence in Five Continents, Vol. 3, IARC Scientific Publ. No. 15, International Agency for Research on Cancer, Lyon, 1976.

36. **Holm, N. V., Hauge, M., and Jensen, O. M.,** Studies of cancer aetiology in a complete twin population: breast cancer, colorectal cancer and leukemia, *Cancer Surv.,* 1, 17, 1982.

37. **Wynder, E. L. and Shigematsu, T.,** Environmental factors of cancers of the colon and rectum, *Cancer,* 20, 1520, 1967.

Chapter 4

DIET AND COLON CANCER: EVIDENCE FROM HUMAN AND ANIMAL MODEL STUDIES

Bandaru S. Reddy

TABLE OF CONTENTS

I. INTRODUCTION

Cancer of the colon is one of the most common tumors observed in the western population, exhibiting more than a tenfold excess when compared to the rural populations in Africa, Asia, and South America.[1-3] During the past 2 decades, epidemiologic studies have investigated the role of environmental factors on the incidence and mortality of colon cancer. Continuing population studies have revealed that our lifestyles, including dietary and nutritional practices, are important variables. These studies also suggested that not only the diets particularly high in total fat and low in certain dietary fibers, vegetables, and micronutrients are generally associated with an increased incidence of colon cancer in man, but dietary fat may be a risk factor in the absence of factors that are protective, such as use of high fibrous foods and fiber.[4-12]

A recent report by Doll and Peto[13] indicate that through dietary modification we might ultimately achieve a reduction of about 90% colon cancer mortality in the U.S. Although the major strength of epidemiologic studies is their focus on human populations, the conduct and interpretation of epidemiologic studies is complicated by inherent problems in testing the dietary practices for their reliability, validity, and sensitivity to reveal narrow but biologically significant differences, and to achieve some degree of dose stratification.[14] However, when another line of evidence based on experimental studies, which have consistently supported human epidemiologic studies supports that diet plays an important role in the etiology of colon cancer, the relationship between diet and colon cancer deserves immediate attention.[7]

The purpose of this review is to evaluate scientific evidence for a dietary etiology of colon cancer and to propose dietary guidelines for the primary and secondary prevention of colon cancer. This review also evaluates various diet-related initiators and promoters involved in colon carcinogenesis and the mechanism whereby dietary factors modulate colon carcinogenesis.

II. EVIDENCE FROM HUMAN NUTRITIONAL EPIDEMIOLOGICAL STUDIES

Cancer of the colon has been the subject of several epidemiologic reviews.[2,3,13] The variability in colon cancer incidence between countries has directed research toward specific environmental dietary factors that are characteristic of high-risk population. While some of these differences may be due to genetic factors or local environmental factors, further evidence for the role of dietary environmental factors in colon cancer has been provided by migrant studies which demonstrate a higher colon cancer incidence rate in the first and second generation Japanese immigrants to the U.S. and in Polish immigrants to Australia than in Japanese and Poles in their respective native countries.[15,16] Furthermore, time-trend in Japan showing that colon cancer seems to be increasing is consistent with the increasing westernization of the Japanese diet.[17] Additional support for dietary factors and lifestyle in the risk for colon cancer came from studies of special religious groups, namely Mormons (members of the Church of Jesus Christ of Latter-Day Saints) and Seventh-Day Adventists (SDA) within the U.S.[18-20] These studies led several investigators to accept diet as a major etiologic factor in colon cancer.

A. Nutritional Factors

1. Role of Dietary Fat and Fat-Containing Foods

Wynder and Shigematsu[21] were the first to suggest that nutritional factors in general and specifically differences in fat intake may be responsible for the international variation in colon cancer incidence. Subsequent descriptive epidemiologic studies have found a strong positive association beween colon cancer mortality or incidence in different countries and per capita availability in national diets of total fat[4,22,23] and of animal fat[23,24] estimated from food balance sheets. Such international correlations may be supportive of a hypothesis, but they should be interpreted with caution because the dietary data were based not on actual intake information but on food disappearance data. The correlation studies were based from populations rather than individuals. In addition, there were differences from country to country in the accuracy with which the data were calculated.

The nutritional epidemiologic studies turned to case-control comparisons and prospective studies in order to accurately define the etiologic factors. Wynder et al.[9] conducted a large-scale retrospective study on colon cancer patients in Japan, which suggested a correlation between the westernization of the Japanese diet and dietary fat and colon cancer. Haenszel et al.[8] demonstrated an association between colon cancer and dietary beef in Hawaiian Japanese cases and Hawaiian Japanese controls, but not in studies in Japan.[25] A recent case-control study in Athens, Greece, demonstrated a positive association between colon cancer and consumption of meat.[26] In another study, no association was found within countries of regional or ethnic colon cancer rates in relation to meat.[27] These conflicting results could be explained on the basis that several of these studies neglected to take into consideration the other confounding factors such as consumption of cruciferous vegetables, dietary fiber, and other food items that have been shown to reduce the risk of colon cancer. In addition, several studies have combined cases of colon and rectal cancer, despite the evidence that these conditions do not have an identical etiology. Finally, some of these studies may have been hampered by the possibility that diets within communities have been too uniform to permit associations between diet and disease to be detected.[28]

A case-control study in Canada indicated an elevated risk for those with an increased intake of calories, total fat, and saturated fat.[29] This study estimated levels of fat consumption by combining information from diet histories with information on the fat content of foods.[29,30] A recent case-control study in Utah Mormons indicated a positive association between dietary fat and colon cancer.[11] These results support a role for total dietary fat in the incidence of colon cancer.

2. Role of Dietary Fiber and Fiber-Containing Foods

Burkitt[6] suggested that countries consuming a diet rich in fiber have a low incidence of colon cancer, whereas those eating refined carbohydrates with little fiber have a higher incidence of the disease. In case-control studies in Israel[31] and in San Francisco area American blacks,[32] it was found that the dietary constituents that were lowest in the diets of patients with colon cancer as compared with controls were those containing high amount of fiber. It is interesting that one of the studies has shown an increased risk of colon cancer to be associated with high-fat/low-fiber intake.[32] Another case-control study showed a lower risk of colon cancer for individuals consuming cruciferous vegetables such as cabbage, broccoli, and Brussels sprouts.[10]

In several populations consuming the diets high in total fat, dietary fiber acts as a protective factor for colon cancer risk. Recent studies comparing populations in rural and urban populations in Finland, Denmark, and Sweden and urban populations in New York indicated that one of the factors contributing to the low risk of colon cancer in rural Scandinavia appears to be high-dietary fiber intake, mainly whole-grain cereals, although all populations are on a high-fat diet.[3,33-35] A strong negative association was reported between regional colon cancer mortality within the U.K. and consumption of fiber foods containing high amount of pentose.[36]

Failure to find consistent strong relationships does not necessarily mediate against a dietary etiology of colon cancer, however, because certain findings may have arisen, at least in part, from methodological limitation of these studies.[28] Inconclusive results of early cor-

Table 1
MODULATING FACTORS AND THEIR MECHANISMS OF ACTION IN COLON CANCER

Dietary fat[a]	Dietary fibers[a,b]	Micronutrients (vitamins, minerals, antioxidants, etc.)
Increases the synthesis of bile acids and their secretion into gut	Certain dietary fibers increase fecal bulk and dilute carcinogens and promoters in the gut	Modify carcinogenesis during the stage of initiation (activation and detoxifcation level) and at promotional phase
Increases metabolic activity of gut microflora	Modify metabolic activity of gut bacteria	
Increases secondary bile acids in colon	Modify the metabolism of carcinogens and/or promotors during the enterohepatic circulation	
Alters immune system		
Stimulation of mixed function oxidase system	Bind the carcinogens and /or promoters and excrete them	

  a   Dietary factors, particularly high total dietary fat and a relative lack of certain dietary fibers and vegetables have a role.
  b   High-dietary fiber, fibrous foods, or cruciferous vegetables may be a protective factor even in the high-dietary fat intake.

relation studies in relation to dietary fiber could be explained by the fact that, in early studies,[23] the statistics of dietary fiber were based on estimates of crude fiber intake, and these calculations underestimate the intake of total dietary fiber.

3. Summary of Nutritional Factors

The bulk of nutritional epidemiologic evidence suggests that diets high in total fat and low in fiber are associated with an increased risk of colon cancer in man (Table 1). In several populations consuming a high amount of total fat, certain dietary fibers and cruciferous vegetables act as protective factors. Moreover, laboratory animal studies discussed elsewhere have clearly demonstrated that high-fat intake promotes and intake of certain fibers inhibits the development of colon cancer. Concurrence between the nutritional epidemiologic and laboratory evidence offered the strength to the concept that diet is a major etiologic factor in colon cancer.

III. MECHANISMS OF NUTRITIONAL FACTORS IN COLON CANCER

The biological plausibility of the nutritional factors in the etiology of colon cancer is reflected in a number of hypotheses for the mechanism of action (Table 1).[37-39] The concept that dietary fat and certain fibers distinct from chemical contaminant of diet and from other environmental and genetic factors are important determinants of colon cancer risk is reinforced by metabolic (biochemical) epidemiologic and laboratory animal studies which have been used to test various hypotheses for the mechanism of action.[40]

Food contains a large number of inhibitors of carcinogenesis, including phenols, indoles, aromatic isothiocyanates, plant sterols, selenium salts, ascorbic acid, tocopherols, and carotenes.[41] These compounds have been shown to inhibit neoplasms in animal models.[41] Since the principal sources of these compounds in the diet are plant constituents, the type and quantity of the plant material in the diet will be of great importance in determining the activity of the protective system. This topic will be discussed in detail in other chapters. Thus, the humans consuming relatively large amounts of vegetables, both cruciferous and noncruciferous type, and fruits would have greater defenses against carcinogens than do individuals consuming a lesser amount of these foods.

The search for genotoxic carcinogens associated with the etiology of colon cancer has been initiated by several laboratories which used Ames *Salmonella typhimurium*/mammalian microsomal assay system to determine the mutagenic (or presumptive carcinogenic) activity in the feces as an end point to understand the nature of these compounds relevant to colon cancer.[42-46] These studies demonstrate that the populations who are at high risk for colon cancer and consuming either high-fat and/or nonvegetarian diets excrete increased levels of fecal mutagens compared to low-risk population who are consuming either a low-fat and/or vegetarian diet.[42-47] The stools of South African urban whites, North American population, and Japanese in Hawaii were higher in mutagenic activity compared to South African urban and rural blacks, Seventh-Day Adventists in North America, vegetarians, Japanese, and Finnish population.[42-47] It has also been shown that increased dietary fiber reduced the fecal mutagens.[43] Although mutagenicity assays of fecal extracts have yet to provide detailed information on the presence of biologically active species in feces, an important finding was that of Hirai et al.[48] who noted that the mutagenic activity of feces of certain donors appears to be due to a type of compound, namely (S)-3-(1,3,5,7,9-dodecapentaenoyloxy)-1,2-propanediol, produced by anaerobic bacteria. However, the carcinogenic activity of this compound is yet to be determined.

Currently, much of our knowledge on the mechanism of dietary fat on colon carcinogenesis is based on experiments conducted in humans (metabolic epidemiology) and animal models.[49] The major significance of these studies is that the primary effect of dietary fat appears to be during the promotional phase of carcinogenesis rather than during initiation phase.[7] The amount of dietary fat modulates the concentration of intestinal bile acids are well as the metabolic activity of gut microflora, which, in turn, metabolize these sterols and other substances into tumorigenic compounds in the colon.[38,39,50] These studies have demonstrated that high-fat diets increase the excretion of bile acids into the gut. These bile acids have been shown to act as colon tumor promoters but do not have the properties of genotoxic carcinogens.[40] This is important since current views on properties of promoters note that the effect of such agents is highly dependent on dose and on length of exposure, and thus provides an opportunity of reducing the risk of colon cancer development by lowering the concentration of bile acids by dietary means.

The mechanism of protective effect of dietary fiber which comprises a heterogenous group of carbohydrates, including cellulose, hemicellulose, and pectin, and a noncarbohydrate substance, lignin, that are resistant to digestion by the secretions of the GI tract has been the subject of a recent workshop.[51] Thus, the protective effect of dietary fiber in colon carcinogenesis depends on the nature and source of fiber in the diet.[52] The protective effect of dietary fiber may be due to adsorption, dilution, and/or metabolism of cocarcinogens, promoters, and yet-to-be identified carcinogens by the components of the fiber. Different types of nonnutritive fiber could bind the tumorigenic compounds, affect the enterohepatic circulation of tumorigenic compounds, as well as dilute potential carcinogens and cocarcinogens by its bulking effect. Thus the humans consuming relatively large amounts of certain fibers would have greater protection against carcinogenic and cocarcinogenic compounds than do individuals consuming lesser quantities of these fibers.

IV. METABOLIC (BIOCHEMICAL) EPIDEMIOLOGIC STUDIES

The concept that dietary fat and certain fibers distinct from chemical contaminants of diet and from other environmental and genetic factors are important determinants of colon cancer risk is reinforced by biochemical epidemiologic studies in humans and laboratory animal studies. A key insight gained from studies in man is that populations who are at high risk for colon cancer development appear to excrete increased amounts of cholesterol and cholesterol metabolites as well as bile acids and mutagens than the populations at low risk. Animal model studies suggest that the dietary factors exert their effect primarily during promotional phase of colon carcinogenssis.

A. Dietary Fat

It has been demonstrated that the concentration of total bile acids and individual bile acids, namely deoxycholic acid and lithocholic acid, is much lower in stools from low-risk populations such as Japanese and other Asians and Africans consuming a low-fat diet when compared to high-risk populations such as North Americans and western Europeans consuming a high-fat diet.[38,50,53] People on a high-fat diet appear to have higher levels of fecal secondary bile acids compared to those on a low-fat diet. In general, there are no major differences in the fecal microflora profiles of these different risk groups although the metabolic activity of some of the constituent microflora, particularly the nuclear dehydrogenating *Clostridia* and bacterial enzymes such as B-glucuronidase and 7α-dehydroxylase, may be associated with the risk for the development of colon cancer. Comparison of fecal bile acids between Seventh-Day Adventists (SDA) consuming a lacto-ovo-vegetarian diet and non-SDAs consuming a high-fat, mixed western diet indicate that fecal excretion of deoxycholic acid and lithocholic acid was lower in SDAs compared with non-SDAs; the fat intake was 28% lower and fiber intake was 1.5 times higher in SDAs, but the total protein and calorie consumption was similar between the groups. It may, therefore, be concluded that the concentration of colonic bile acids has a major role in determining the risk of developing colon cancer.

B. Dietary Fiber

Reddy et al.,[33,35] Domellof et al.,[34] and Jensen et al.[54] have demonstrated that the concentration of fecal secondary bile acids, deoxycholic acid and lithocholic acid, was lower in populations of rural Scandinavia who are at low or intermediate risk for colon cancer and consuming high-fat/high-fiber diets than in urban Copenhagen (Denmark), Malmo (Sweden), and metropolitan New York (Table 2). The dietary histories indicate that the total fat and protein intake in high- and low-risk areas is similar, but the consumption of fiber, mainly cereal type, is 1.5 to 2.5 times higher in low-risk areas. The total dietary fiber intake in Kuopio, Umea, Malmo, and New York are 32, 26, 17, and 14 g/day, respectively. The daily output of bile acids remained the same in all population groups because the dietary intake of fat is the same in these areas. The concentration of bile acids was lower in populations consuming high-fat/high-fiber diets because of the larger stool bulk than that of populations consuming high-fat/low-fiber diets.

V. EVIDENCE FROM LABORATORY ANIMAL STUDIES

A number of distinct animal models of colon cancer, induced by chemicals and operating by different metabolic and biochemical mechanisms, are available for studying the pathogenesis of colon cancer and comparing it to similar stages of the disease seen in man.[55,56] Additionally, these animal models have been used as unique tools for systematic studies of the risk factors observed in human setting and for determining whether or not suspected etiologic factors can be reproduced under controlled laboratory conditions. Thus, as is described here, a number of major elements observed in humans, such as the enhancing effect on colon carcinogenseis due to fat or other protective effect of certain dietary fibers or micronutrients, could not have been established without careful, deliberate investigations carried out in laboratory animals.

Table 2
FECAL BILE ACIDS OF HEALTHY SUBJECTS FROM MALMO (SWEDEN), UMEA, SWEDEN, KUOPIO (FINLAND), AND METROPOLITAN NEW YORK[33-35]

	mg/g dry feces			
Bile acids	Kuopio	Umea	Malmo	New York
Cholic acid	0.20 ± 0.06	0.3 ± 0.13	0.24 ± 0.15[a]	0.2 ± 0.04
Chenodeoxycholic acid	0.13 ± 0.03	0.2 ± 0.05	0.22 ± 0.05	0.2 ± 0.03
Deoxycholic acid	1.72 ± 0.16[b]	2.2 ± 0.55[b]	4.02 ± 0.42	4.5 ± 0.30
Lithocholic acid	1.40 ± 0.16[b]	2.3 ± 0.38[b]	3.94 ± 0.32	4.0 ± 0.20
Ursodeoxycholic acid	0.08 ± 0.02[b]	0.4 ± 0.13	0.53 ± 0.13	0.3 ± 0.04
$3\alpha,7\beta,12\alpha$-Trihydroxy-5β-cholanic acid	0.04 ± 0.01[b]	0.3 ± 0.12	0.25 ± 0.09	0.2 ± 0.04
12-ketolithocholic acid	0.06 ± 0.02[b]	0.3 ± 0.08	0.41 ± 0.10	0.4 ± 0.02
Other bile acids	0.93 ± 0.08	1.7 ± 0.31	1.42 ± 0.18	3.3 ± 0.40
Total bile acids	4.59 ± 0.42[b]	7.7 ± 1.10[b]	11.02 ± 1.90	13.1 ± 0.60

[a] Values are means $\pm$ SE.
[b] Significantly different from New York and Malmo, $p > 0.05$.

A. Dietary Fat

The possible role of dietary fat on colon carcinogenesis has received support from studies in animal models. In several earlier studies on dietary fat and colon cancer, interpretation of results between high- and low-fat diets was complicated by the use of diets of varying caloric density and confounded by different intakes of other nutrients. However, recent studies in which the intake of all nutrients and total calories were controlled between the high- and low-fat groups, indicated that the amount of dietary fat is an important factor in colon carcinogenesis.[40]

The stage of carcinogenesis at which the effect of dietary fat is exerted appears to be during the promotional phase of carcinogenesis, rather than during initiation phase.[40] Ingestion of high beef fat increased the intestinal tumor incidence when fed after azoxymethane (AOM) treatment, but not during or before the carcinogen administration.[59] A recent study indicates that the dietary unsaturated fat alters the metabolism of 1,2-dimethylhydrazine (DMH) and thus influences the carcinogenic process during the initiation phase of carcinogenesis.[60] However, in this study, the rats were fed the experimental high-fat diet before, during, and after the carcinogen treatment, making it difficult to distinguish whether the effect of fat is at the initiation or during the postinitiation stage of carcinogenesis.

Definitive pharmacokinetic and metabolic data are not available to give a complete view of the influence of dietary fat on the metabolism of colon carcinogenens. Therefore, studies are warranted to examine in detail the effects of types and amounts of dietary fat on several key processes of colon carcinogen activation in the liver and colonic mucosa.

1. Amount

Investigations were also carried out to determine the effect of high dietary fat on colon tumor induction by a variety of carcinogens, DMH, MAM acetate, 3,2'-dimethyl-4-aminobiphenyl (DMAB), methylnitrosourea (MNU), which not only differ in metabolic activation, but also represent a broad spectrum of exogenous carcinogens.[57,58] Male F344 weanling rats were fed semipurified diets containing 20 or 5% beef fat. At 7 weeks of age, animals were given DMH (s.c., 150 mg/kg body weight, one dose), MAM acetate (i.p., 35 mg/kg body weight, one dose), DMAB (s.c., 50 mg/kg body weight weekly for 20 weeks), or

Table 3
EFFECT OF TYPE AND AMOUNT OF DIETARY FAT
ON COLON TUMORS IN F344 RATS

Experiment no.	Type of fat	%Dietary fat	Carcinogen	% Rats with colon tumors
1	Lard	5	DMH[a]	17
		20	DMH	67
	Corn oil	5	DMH[a]	36
		20	DMH	64
2	Beef fat	5	DMH[b]	27
		20	DHM	60
		5	MNU[c]	33
		20	MNU	73
		5	MAM acetate[d]	45
		20	MAM acetate	80
3	Beef fat	5	DMAB[e]	26
		20	DMAB	74
4	Corn oil	5	AOM[f]	17
		20	AOM	46
	Safflower oil	5	AOM	13
		20	AOM	36
	Olive oil	5	AOM	10
	Olive oil	5	AOM	10
		20	AOM	13
	Coconut oil	20	AOM	13

[a] Female F344 rats, at 4 weeks of age, were given DMH s.c. at a weekly dose rate of 10 mg /kg body weight for 20 weeks and autopsied 10 weeks later.

[b] Male F344 rats, at 7 weeks of age, were given a single s.c. dose of DMH, 150 mg/kg body weight and autopsied 30 weeks later.

[c] Male F344 rats, at 7 weeks of age, were given MNU i.r., 2.5 mg per rat, twice a week for 2 weeks and autopsied 30 weeks later.

[d] Male F344 rats, at 7 weeks od age, were given a single i.p. dose of MAM acetate, 35 mg /kg body weight and autopsied 30 weeks later.

[e] Male F344 rats, at 7 weeks of age, were given DMAB s.c. at a weekly dose rate of 50 mg/kg body wt for 20 weeks and autopsied 20 weeks later.

[f] Female F344 rats, at 5 weeks of age, were fed low-fat diets. At 7 weeks of age, they were given a single dose of AOM. 20 mg/kg body weight, s.c. One week later, they were transferred to their respective high-fat diet and autopsied 48 weeks later.

MNU (intrarectal, 2.5 mg/rat weekly for 2 weeks) and autopsied 30 to 35 weeks later. Irrespective of the colon carcinogens that differed in metabolic activation, animals fed a diet containing 20% beef fat had a greater incidence of colon tumors than did rats fed a diet containing 5% beef fat (Table 3).

Nigro et al.[61] induced intestinal tumors in male Sprague-Dawley rats by s.c. administration of AOM and compared animals fed Purina chow with 35% beef fat to those fed Purina chow containing 5% fat. Animals fed the high fat developed more intestinal tumors and more metastasis into abdominal cavity, lungs, and liver than the rats fed the low-fat diet. In another study, W/Fu rats fed a 30% lard diet had an increased number of DMH-induced large-bowel tumors compared to the animals fed the standard diet.[62] Rogers and Newberne[63] found that a diet marginally deficient in lipotropes but high in fat, enchanced DMH-induced colon carcinogenesis in Sprague-Dawley rats. Diets containing 20% corn oil or 20% safflower oil markedly increased the AOM-induced colon tumors in F344 rats as compared with diets containing 5% corn oil or 5% safflower oil.[64] However, a recent study indicates that a high-fat diet containing 24% corn oil, 24% beef fat or 24% Crisco had no effect on colon tumors induced by oral doses of DMH in Sprague-Dawley rats.[65] However, in this study,[65] all diets were prepared in a 1:1 ratio into a 5% agar solution and contained slightly higher levels of minerals and vitamins over recommended levels. Thus, it is difficult to interpret the results. It has been shown that dietary agar and a related compound, carrageenan, enhance colon tumor evoked by DMH, AOM, or MNU.[66,67] These results thus suggest that total dietary fat may have a function in the pathogenesis colon cancer.

2. Type

Reddy et al.[68] designed experiments to study the effect of a particular type and amount of dietary fat for two generations before animals were exposed to treatment with a carcinogen. F344 rats fed 20% lard or 20% corn oil were more susceptible to colon tumor induction by DMH than those fed 5% lard or 5% corn oil. The type of fat appears to be immaterial at the 20% level, although at the 5% fat level there is a suggestion that unsaturated fat (corn oil) predisposes to more DMH-induced colon tumors than saturated fat (lard). Sprague-Dawley rats fed a 20% safflower oil diet[69] or 10% corn oil diet[60] had more DMH-induced colon tumors than did those fed a 20% coconut oil or a 9% coconut oil + 1% linoleic acid diet, respectively. Donrye rats fed a semipurified diet containing 5% linoleic acid demonstrated a significantly higher incidence of colon tumors, more tumors per rat, and greater malignant differentiation than did those 4.7% stearic acid + 0.3% essential fatty acid.[70]

A recent study from our laboratory in which intake of all nutrients and calories except fat calories were controlled provides some evidence for the effect of type of fat on colon carcinogenesis.[64] AOM-induced colon tumor incidence was increased in F344 rats fed 20% corn oil or 20% safflower oil diets compared to those fed 5% corn oil or 5% safflower oil diets (Table 4). However, diets containing 20% coconut oil or 20% olive oil had no enhancing effect on colon tumor promotion; rats fed the 20% olive oil diet or a 20% coconut oil diet had a colon tumor incidence the same as that in rats fed the diets containing 5% olive oil, 5% corn oil, or 5% safflower oil. In summary, these and other results indicate that diets high in saturated fat of vegetable origin (coconut oil) or monounsaturated fat of vegetable origin (olive oil), polyunsaturated fats (corn oil or safflower oil), and saturated fats of animal origin (lard or beef tallow) differ in colon tumor promotion. The varied effects of different types of fat on colon cancer suggest that the fatty acid composition is one of the important factors in determining the modifying effect of various fats in colon tumor promotion.

B. Dietary Fiber

The relation between dietary fiber and colon cancer has been the subject of a recent workshop.[51] Although the concept of dietary fiber involvement in colon carcinogenesis is of great importance, studies examining the possible role of various types of dietary fiber in animal models appear to have provided conflicting results. This discrepancy might have been, in part, due to the nature of the carcinogen used, to differences in the susceptibility of rat strain to carcinogen treatment, to variation in the composition of diets, to qualitative and quantitative differences in administered intact fibers and their components, to relative differences in food intake by the animals and/or to differences in experimental design and duration of the experiment.

In an earlier study, Cruse et al.[71] found that a diet containing 20% wheat bran had no inhibitory effect on colon carcinogenesis induced by DMH in rats. However, the doses of DMH in this experiment were so high that any protective effect of bran might have been unobservable. It is important to avoid exposing the animal to an excessive level of carcinogen for a long period, as this may obscure more subtle changes induced by certain dietary modifications. In fact, the data presented by Cruse and associates[71] suggest that a high-fiber

Table 4
AOM-INDUCED INTESTINAL TUMOR INCIDENCES IN FEMALE F344 RATS FED THE DIETS CONTAINING CORN OIL, SAFFLOWER OIL, OLIVE OIL, COCONUT OIL, AND MEDIUM CHAIN TRIGLYCERIDES

| | %Animals with tumors of | | | | | | Colon tumors/animal | | |
| | Colon | | | Small intestine | | | | | |
Diet group	Total	Adenoma	Adenocarcinoma	Total	Adenoma	Adenocarcinoma	Total	Adenoma	Adenocarcinoma
Low corn oil (30)[a]	17	10	7	3	0	3	0.17[b]	0.10	0.07
High corn oil (28)	46[c]	14	32	0	0	0	0.46	0.14	0.32
Low safflower oil (30)	13	7	6	3	0	3	0.13	0.07	0.06
High safflower oil (28)	36[c]	14	22	4	0	4	0.36	0.14	0.22
Low olive oil (29)	10	7	3	10	0	10	0.10	0.07	0.03
High olive oil (30)	13	10	3	7	0	7	0.13	0.10	0.03
High coconut oil	13	10	3	0	0	0	0.13	0.13	0.0

[a]   Effective number of animals are shown in parenthesis.

[b]   Values are expressed as means.

[c]   Significantly different from their respective 5% fat diet at $p < 0.05$.

Table 5

**COLON TUMOR INCIDENCE IN F344 MALE RATS FED DIETS CONTAINING
WHEAT BRAN OR CITRUS FIBER AND TREATED WITH AZOXYMETHANE**

Diet	%Animals with colon tumors			Colon tumors per tumor-bearing rat		
	Total[a]	Adenoma	Adenocarcinoma	Total	Adenoma	Adenocarcinoma
Control	90	86	63	3.45 ± 0.16[b]	2.37 ± 0.16	1.08 ± 0.18
Wheat bran	71[c]	47[c]	39[c]	1.55 ± 0.12[d]	0.94 ± 0.13[d]	0.61 ± 0.11[d]
Citrus pulp	63[c]	41[c]	39[c]	1.78 ± 0.18[d]	0.90 ± 0.14[d]	0.88 ± 0.16

[a]   Total represents animals with adenomas and/or adenocarcinomas.
[b]   Mean ± SEM.
[c]   Significantly different from the group fed the control diet by X^2 test ($p < 0.05$).
[d]   Significantly different from the group fed the control diet by Student's t-test ($p < 0.05$).

diet reduces the frequency of death due to DMH in rats. Sprague-Dawley rats fed a diet containing 20% corn oil or beef fat and 20% wheat bran had fewer DMH-induced colon tumors than those on a control diet containing 20% fat and no bran.[72] In another study, addition of 10% wheat bran to the 35% beef fat diet had no effect on AOM-induced colon tumor frequency in Sprague-Dawley rats.[73] Apparently, the AOM dosage and the amount of fat were too great to be affected by the dietary fiber.[73] However, in the same study, addition of 20% wheat bran or cellulose, but not 20% alfalfa, to a diet containing 6% fat significantly reduced the intestinal tumor frequency as well as tumor frequency in the proximal colon.[73] In a study by Bauer et al.,[74] groups of rats were fed a fiber-free diet or diets containing 20% wheat bran from 3 days prior to the first DMH injection until 14 days after the last injection. They were then transferred to standard rat pellet-diet for about 10 to 12 weeks before sacrifice. There was no difference in the incidence of colorectal tumors among the dietary treatments. In another study,[74] colon adenomas and adenocarcinomas were increased in rats fed a 20% wheat bran diet during the DMH treatment compared to those fed a fiber-free diet. There was, however, inhibition of colon tumors in rats fed the 20% wheat bran diet after the carcinogen treatment. It is possible in these studies[74,75] that a switch from a high-fiber experimental diet to a standard diet or a fiber-free diet during the postcarcinogen period failed to show any protective effect of wheat bran. It is also possible that a high wheat bran (20%) diet, when fed during the carcinogen administration, caused unavailability of certain essential trace metals that are involved in carcinogen detoxification, thereby enhancing colon tumorigenesis.

The effect of a diet containing 15% alfalfa or wheat bran on colon carcinogenesis by MNU or AOM was studied in F344 rats by Watanabe et al.[76] The animals fed the 15% alfalfa diet and treated with MNU had a higher incidence of colon tumors than those fed a control diet containing only 5% cellulose or 15% wheat bran. The incidence of colon tumors induced by AOM in rats fed a diet containing wheat bran was lower than that in rats fed a control diet or the alfalfa diet. The effect of dietary wheat bran (15%) and dehydrated citrus fiber (15%) on AOM or DMAB-induced intestinal tumors was studied in male F344 rats (Table 5).[77,78] Animals fed the wheat bran and citrus fiber diets and treated with AOM had a lower incidence of colon and small intestinal tumors. Other studies have shown that dietary corn bran (15%) enhanced colon tumor incidence in rats.[79,80]

These studies thus indicate that the diets containing wheat bran and citrus fiber reduce the risk of chemically induced colon cancer in animal models and that the protection against colon cancer depends on the type of intact fiber. The varied effects of different fiber sources on colon carcinogenesis may be due to their soluble and insoluble components, i.e., cellulose, hemicellulose, lignin, and pectin. It is possible that the chemical composition, in addition to the physical properties of dietary fiber, is important in determining the modulating effect of various fibers on colon carcinogenesis.

The relationship between the components of dietary fiber and colon cancer has also been studied in animal models. Freeman[81] compared the incidence of colon tumors induced by DMH in Sprague-Dawley rats fed either a fiber-free diet or a diet containing 4.5% purified cellulose or pectin. Among the animals ingesting cellulose, fewer had colonic neoplasms, and the total number of colon tumors in this group was lower. The effect of a diet containing 15% pectin on colon carcinogenesis by MNU or AOM was studied in F344 rats by Watanabe et al.[76] The addition of pectin to the diet greatly inhibited colon tumor incidences induced by AOM, a carcinogen requiring metabolic activation, but not induced by MNU, a direct-acting carcinogen. In a recent study, Reddy et al.[80] showed that the incidence and multiplicity of small intestinal tumors as well as the multiplicity of colon adenocarcinomas induced by DMAB were lower in F344 rats fed the 7.5% lignin diet than in those fed the control diet containing 5% cellulose. These results thus indicate that the protective effect of fiber components in colon carcinogenesis depends on the type of carcinogen and the types of fiber components.

C. Dietary Fat and Fiber and Bile Acid Excretion

In order to understand the specifics of the mechanisms whereby dietary fat influences colon cancer, the effect of type and amount of dietary fat on biliary and fecal bile acids was studied in rats.[64,82,83] These studies indicate that the effect of dietary fat on colon carcinogenesis is mediated through a concomitant change in the concentration of colonic bile acids. Biliary excretion of cholic acid, β-muricholic acid, and deoxycholic acid was higher in rats fed diets containing 20% corn oil or 20% lard than in rats fed diets containing 5% corn oil or 5% lard. High-fat (corn oil or lard) intake was associated with an increased excretion of fecal secondary bile acids: deoxycholic acid and lithocholic acid. Type of fat (corn oil vs. lard) had no effect on biliary and fecal bile acid excretion. In another study, there was a significant increase in the concentration of colonic deoxycholic acid and lithocholic acid in rats fed diets containing 20% corn oil or 20% safflower oil when compared to rats fed diets containing 5% corn oil or 5% safflower oil.[64] In contrast, there was no difference in the concentrations of deoxycholic acid and lithocholic acid between animals fed the 20% olive oil diet, 5% olive oil diet, or 20% coconut oil diet (Table 6).

The effect of types and components of dietary fiber on fecal bile acid excretion was investigated in rats.[80,84] The concentrations of fecal bile acids, particularly hyodeoxycholic acid, deoxycholic acid, and lithocholic acid were lower in rats fed wheat bran compared to those fed a control diet without wheat bran. Dietary corn bran increased the concentration of fecal deoxycholic acid as compared to the level found with the control diet without corn bran.

Thus, the excretory pattern of fecal secondary bile acids observed in these studies correlated with colon tumor incidences in animal models. These studies also suggest that high dietary intake of certain types of fat and low intake of certain fibers may be necessary for the full expression of risk for colon cancer.

D. Bile Acids and Colon Tumor Promotion

The role of nutritional influences through the bile acids as promoting stimuli has been noted for the colon and will be discussed with experimental evidence. Taking a lead from human metabolic epidemiologic information, a test system was devised in rats to demonstrate the effect of various bile metabolites in colon carcinogenesis.

The evidence of the importance of bile acids as colon tumor promoters came from our studies.[86,87] Lithocholic acid or deoxycholic acid applied topically to the colon increased

Table 6
FECAL BILE ACID EXCRETION IN VEHICLE-TREATED FEMALE F344 RATS FED THE SEMIPURIFIED DIETS CONTAINING VARIOUS FATS

Dietary group	Cholic acid	Chenodeoxycholic acid	β-Muricholic acid	Hyodeoxycholic acid	Lithocholic acid	Deoxycholic acid	12-Ketolitho-cholic acid	Other bile acids	Total bile acids
				mg/g dry feces					
5% Corn oil	0.11 ± 0.04[a,b,1]	0.03 ± 0.01[1]	0.26 ± 0.05[1]	0.84 ± 0.06[1,2]	0.30 ± 0.09[1]	1.20 ± 0.10[1]	0.09 ± 0.01[1]	0.68 ± 0.1	3.53 ± 0.4[1]
20% Corn oil	0.14 ± 0.05[1]	0.02 ± 0.01[1]	0.29 ± 0.06[1]	0.98 ± 0.09[1,2]	0.68 ± 0.07[2]	2.68 ± 0.26[2]	0.20 ± 0.02[2]	1.18 ± 0.3	6.17 ± 0.7[2]
5% Safflower oil	0.14 ± 0.06[1]	0.07 ± 0.02[1]	0.28 ± 0.08[1]	0.70 ± 0.08[1]	0.24 ± 0.04[1]	1.10 ± 0.14[1]	0.06 ± 0.01[1]	0.86 ± 0.2	3.45 ± 0.5[1]
20% Safflower oil	0.18 ± 0.08[1]	0.05 ± 0.02[1]	0.33 ± 0.06[1]	1.20 ± 0.09[2]	0.42 ± 0.04[2]	2.98 ± 0.3[2]	0.31 ± 0.04[2]	0.98 ± 0.2	6.45 ± 0.7[2]
5% Olive oil	0.12 ± 0.05[1]	0.09 ± 0.02[1]	0.11 ± 0.06[2]	0.96 ± 0.10[1,2]	0.28 ± 0.08[1]	1.01 ± 0.15[1]	0.14 ± 0.04[1]	0.81 ± 0.3	3.42 ± 0.6[1]
20% Olive oil	0.38 ± 0.04[2]	0.19 ± 0.01[2]	0.18 ± 0.04[2]	1.30 ± 0.28[2]	0.36 ± 0.1[1]	1.30 ± 0.20[1]	0.19 ± 0.05[1]	1.20 ± 0.3	5.10 ± 0.5[3]
20% Coconut oil	0.26 ± 0.05[2]	1.12 ± 0.02[1,2]	0.56 ± 0.08[3]	1.35 ± 0.26[2]	0.52 ± 0.05[2]	1.30 ± 0.24[1]	0.16 ± 0.02[1]	0.99 ± 0.4	5.26 ± 0.8[3]

[a]   Values are expressed as means ± SEM (n = 6).

[b]   Means in the same column which do not share a common superscript number are significantly different at $p < 0.05$ (Duncan's Multiple Range test).

MNNG-induced colon adenocarcinomas in rats. The bile acids themselves did not produce any tumors. Cohen et al.[88] reported that cholic acid in the diet increased MNNG-induced colon carcinogenesis in rats. Total fecal bile acids, particularly deoxycholic acid output, were elevated in animals fed cholic acid as compared with controls. This increase in fecal deoxycholic acid was due to bacterial 7α-dehydroxylation of cholic acid in the colonic contents. These studies demonstrate that the the secondary bile acids have a promoting effect in colon carcinogenesis.

Although the molecular mechanisms of tumor-promoting action of bile acids are incompletely understood, recent studies of Takano et al.[89] suggest that the induction of colonic epithelial ornithine decarboxylase activity and bile acid administration may play a role in these mechanisms. Although the classical tumor promoter, TPA, showed more potent induction of ornithine decarboxylase activity than the deoxycholate, the maximal induction was greater in the case of deoxycholate treatment. The studies reported indicate that the colonic epithelial responses of the polyamine biosynthetic enzymes to applications of bile acids are among the earliest changes to occur in this tissue in response to promoting agents. This study also indicates that a specific bile acid structure with a definite spatial relationship of the hydroxyl groups (lithocholate, chenodeoxycholate, deoxycholate, apocholate, etc.) is required for induction of ornithine decarboxylase activity.

Bile acids have been shown to affect cell kinetics in the intestinal epithelium, although the structural specificity of this effect has not been examined extensively. Recently, Cohen et al.[88] reported an enhanced colonic cell proliferation in rats fed cholic acid, as well as in animals treated with intrarectal MNU. This increased cell proliferation involved in DNA synthesis by cholic acid feeding would favor the expression of damage at a far higher level than with the carcinogen MNU alone, bringing about not only a greater overall incidence of MNU-induced colon acid. Takano et al.[89] reported that deoxycholate administered intrarectally stimulated colonic epithelial DNA synthesis. Lipkin[90] demonstrated that, during neoplastic transformation of colonic cells, a similar sequence of changes leading to uncontrolled proliferative activity develops in colon cancer in humans and in rodents given a colon carcinogen. Irrespective of the mechanism by which bile acids enhance cell proliferation and (or) decrease the generation time of proliferating cells and affect DNA repair, the phenomenon may have important implications for colon carcinogenesis. Obviously, further studies are warranted on the mechanism of tumor-promoting activity of various bile acids.

VI. PROSPECTS FOR PRIMARY AND SECONDARY PREVENTION OF COLON CANCER

What can be done to reduce mortality from colon cancer by taking advantage of currect knowledge? What other kinds of information need to be generated to promote control of colon cancer?

During the last decade, a substantial amount of progress has been made in the understanding of the relationship between the dietary constituents and the development of colon cancer in man. The information base is sufficiently convincing with respect to an enhancing effect as a function of total fat intake and a protective effect of certain dietary fibers in colon cancer. The populations with high incidence of colon cancer are characterized by consumption of high-dietary fat which may be a risk factor in the absence of factors that are protective, such as use of whole-grain cereals, high fibrous foods, and vegetables mainly of cruciferous type. Application of the findings made thus far in colon cancer research for the general public is, therefore, to have a far-reaching impact on the major premature, killing diseases in the western world.

Current research in animal models which suggests that the carcinogenic response to a variety of colon carcinogens is enhanced by the dietary fat and inhibited by several dietary fibers, indicates that these nutritional factors may operate during the promotional phase of carcinogenesis. The carcinogenic process in humans may have similar characteristics. The fact that ubiquitous environmental carcinogens are present at very low concentrations and the extent of the carcinogenic stress from this source is probably rather weak suggests that promoting factors may have a preponderant influence on the eventual outcome of the neoplastic process in humans. The understanding of postinitiating events appears to offer some promise that intervention in the cancer process in man prior to the occurrence of overt tumors may be an achievable and realistic goal. Because promotion or postinitiating events are a reversible process, in contrast to the rapid irreversible or long-lasting process of initiation by carcinogens, manipulation of promotion would seem to be the ideal method of colon cancer prevention. However, in prevention, it makes little difference by what mechanism an agent operates, provided that its partial or total elimination can be shown to lead to a decline in cancer incidence. In this regard, advice to the public at large assumes particular importance because several decades may span the gap between initiating and clinical manifestation of cancer, and, therefore, steps taken today may have a major impact on the nature of future events.[91,92]

If the hypothesis that the same factors that affect colon cancer incidence also influence the recurrence of adenomatous polyps after polypectomy, transformation of adenomas into carcinomas of the colon, and recurrence of colon carcinomas after surgical intervention in cancer patients is correct, then dietary intervention in these patients should result in objective increase in disease-free survival and overall survival of these patients. The same thing holds good also to the symptomatic and asymptomatic patients affected with autosomal dominantly inherited nonpolyposis colorectal cancer syndromes which may account for as much as 5 to 7% of all occurences of colorectal cancer. The overall goal is to reduce the incidence of animal models, but also will question about participant acceptance of the intervention.[93]

These prudent measures for the general population as well as patients or families at high risk for colon cancer development are imperative when various lines of evidence is concordant and points to current dietary patterns of the U.S. and western countries with an elevated risk for colon cancer. The advantage of encouraging the public to alter their dietary pattern, stemming from combined knowledge obtained from the industrialized, developed countries where a lower fat intake as well as high dietary fiber yield a lower risk from colon cancer, is that these measures have no obvious adverse effects as is witnessed by the general health of the public in these countries. Therefore, these dietary recommendations can be made without concern for incurring any adverse effects as might be found with a drug-oriented management and disease prevention program. Therefore, in terms of prevention of colon cancer, a decision to alter dietary habits leading to a lower intake of dietary fat and higher intake of certain dietary fibers has potential beneficial effects which go beyond colon cancer in that a reduction in fat intake might also influence the risk for other diet-related cancers and, in particular, could reduce the rate of coronary heart disease.

VII. DIETARY RECOMMENDATIONS

The significance of the proposed dietary recommendations lies in its ability to provide information relevant to the use of dietary intervention as a form of totally effective nontoxic therapy for patients with colon cancer and adenomatous polyps. These recommendations apply to the public at large.

The modified dietary regimen consists of a reduction of dietary fat to 25% of fat calories with polyunsaturated, saturated, and monounsaturated fatty acid ratio to 1:1:1 and an increase in complex carbohydrate intake. Total dietary fiber intake per day shall be increased to 32 g based on McCance and Widdowson's fiber values[94] which is aproximately 25 g of neutral detergent fiber. It has been shown that a moderate reduction of the dietary fat intake to 20 to 25% of total calories does not appear to constitute a health risk if an adequate ratio of polyunsaturated:saturated fatty acids is maintained.[95] There is concern about the fat-soluble vitamin intake since the level of intake of fat would result in the intake of fat-soluble vitamins, mainly vitamin A. The fatty foods are replaced by those rich in complex carbohydrates from products of vegetable origin which are a good source of vitamin A precursors, the carotenes. There is also a concern that high-fiber diets have some effect on the availability of certain nutrients, particularly minerals, in the body. Diets containing less than 25 g of neutral detergent fiber and fat at 20% or more of the total calories can be considered safe.[92]

In order to achieve the above objective, we recommend eating fewer foods high in saturated and unsaturated fats and to increase the consumption of fruits, vegetables, and whole-grain cereal products daily. These fruits and vegetables should include oranges, grapefruits, apples, dark green leafy vegetables, carrots, winter squash, tomatoes, cabbage, cauliflower, and Brussels sprouts, to cite a few. This is not a diet which we generally adopt as an emergency measure after we have indulged in nutritional excesses, but rather a continuing life-long, low-risk, yet pleasant, tasteful, and nutritious diet from the public health point of view. Therefore, the time to make nutritional declarations to western populations is now. The oncologists will have the opportunity to monitor the success of such nutritional modification.

REFERENCES

1. **Waterhouse, J., Muir, C., Correa, P., and Powel, J.,** Cancer Incidence in Five Continents, Vol. 3, IARC Sci. Publ. No. 15, International Agency for Research on Cancer, Lyon, France, 1976.
2. **Correa, P. and Haenszel, W.,** The epidemiology of large bowel cancer, *Adv. Cancer Res.,* 26, 1, 1978.
3. **Jensen, O. M.,** Colon cancer epidemiology, in *Experimental Colon Carcingenesis,* Autrup, H. and Williams, G. M., Eds., CRC Press, Boca Raton, Fla., 1983, 3.
4. **Armstrong, D. and Doll, R.,** Environmental factors and cancer incidence and mortality in different countries, with special reference to dietary practices, *Int. J. Cancer,* 15, 617, 1975.
5. National Research Council, Diet, Nutrition and Cancer, Assembly of Life Sciences, National Research Council, National Academy Press, Washington, D.C., 1982.
6. **Burkitt, D. P.,** Colonic-rectal cancer: fiber and other dietary factors, *Am. J. Clin. Nutr.,* 31, S58, 1978.
7. **Reddy, B. S., Cohen, L. A., McCoy, G. D., Hill, P., Weisburger, J. H., and Wynder, E. L.,** Nutrition and its relationship to cancer, *Adv. Cancer Res.,* 32, 237, 1980.
8. **Haenszel, W., Berg, J. W., Segi, M., Kurihara, M., and Locke, F. B.,** Large bowel cancer in Hawaiian Japanese, *J. Natl. Cancer Inst.,* 51, 1765, 1973.
9. **Wynder, E. L., Kajitani, T., Ishikawa, S., Dodo, H., and Takano, A.,** Environmental factors of the colon and rectum. II. Japanese epidemiological data, *Cancer,* 23, 1219, 1969.
10. **Graham, S. and Mettlin, C.,** Diet and colon cancer, *Am. J. Epidemiol.,* 109, 1, 1979.
11. **West, D. K., Lyon, J. L., Gardner, J. W., Schuman, K., Stanish, W., Mahoney, A., Sorenson, A., and Avlon, E.,** Epidemiology of colon cancer in Utah, in 1983 Workshop on a Decade of Achievements and Challenges in Large Bowel Cancer Research 1972—1982, National Large Bowel Cancer Project, Houston, Tex., 1983, 3.
12. **Jensen, O. M.,** Epidemiological evidence associating lipids with cancer causation, in *Dietary Fats and Health,* Perkins, E. G. and Visek, W. J., Eds., American Oil Chemists' Society, Champaign, Ill., 1983, 698.
13. **Doll, R. and Peto, R.,** The causes of cancer: quantitative estimates of avoidable risks of cancer in the United States today, *J. Natl. Cancer Inst.,* 66, 1192, 1981.
14. **Schottenfeld, D.,** Overview, in *Current Topics in Nutrition and Disease,* Vol. 9, Roe, D. A., Ed., Alan R. Liss, New York, 1983, 197.
15. **Haenszel, W., Berg, J. W., Segi, M., Kurihara, M., and Locke, F. B.,** Large bowel cancer in Hawaiian Japanese, *J. Natl. Cancer Inst.,* 51, 1765, 1973.
16. **Staszewski, J., McCall, M. G., and Stenhouse, N. S.,** Cancer mortaility in 1962—66 among Polish migrants to Australia, *Br. J. Cancer,* 25, 599, 1971.
17. **Hirayama, T.,** Diet and cancer, *Nutr. Cancer,* 1, 67, 1979.

18. **Phillips, R. L., Garfinkel, L., Kuzma, J. W., Beeson, W. L., Lotz, T., and Brin, B.,** Mortality among California Seventh day Adventists for selected cancer sites, *J. Natl. Cancer Inst.*, 65, 1097, 1980.
19. **Lyon, J. L., Gardner, J. W., and West, D. W.,** Cancer risk and lifestyle: cancer amoung Mormons from 1967—1975, in *Cancer Incidence in Defined Populations*, Banbury Report No. 4, Cairns, J., Lyon, J. L., and Skolnick, M., Eds., Cold Spring Harbor Laboratory, New York, 1980, 3.
20. **West, D. K., Lyon, J. L., Gardner, J. W., Schwan, K., Stanish, W., Mahoney, A., Sorenson, A., and Avlon, E.,** in *Cancer Incidence in Defined Populations*, Banbury Report No. 4, Cairns, J., Lyon, J. L., and Skolnick, M., Eds., Cold Spring Harbor Laboratory, Cold Spring Harbor, New York, 1980, 31.
21. **Wynder, E. L. and Shigematsu, T.,** Environmental factors of cancers of the colon and rectum, *Cancer*, 20, 1520, 1967.
22. **Carroll, K. K. and Khor, H. T.,** Dietary fat in relation to tumorigenesis, *Prog. Biochem. Pharmacol.*, 10, 308, 1975.
23. **Drasar, B. S. and Irving, D.,** Environmental factors in cancer of the colon and breast, *Br. J. Cancer*, 27, 167, 1973.
24. **Howel, M. A.,** Diet as an etiological factor in the development of cancers of the colon and rectum, *J. Chronic Dis.*, 28, 67, 1975.
25. **Haenszel, W., Locke, F. B., and Segi, M.,** A case-control study of large bowel cancer in Japan, *J. Natl. Cancer Inst.*, 64, 17, 1980.
26. **Manousos, O., Day, N. E., Trichopoulous, D., Gervassilis, E., Tzonow, A., and Polychronopoulous, A.,** Diet and colorectal cancer: a case-control study in Greece, *Int. J. Cancer*, 32, 1, 1983.
27. **Kolonel, L. N., Hankin, J. H., Lee, J., Chu, S. Y., Nomura, A. M. Y., and Ward, H. M.,** Nutrient intakes in relation to cancer incidence in Hawaii, *Br. J. Cancer*, 44, 332, 1981.
28. **McKeown-Eyssen, G.,** A diet to prevent colon cancer: how do we get there from here?, in *Diet, Nutrition, and Cancer: From Basic Research*, Roe, D. A., Ed., Alan R. Liss, New York, 1983, 243.
29. **Jain, M., Cook, G. M., Davis, F. G., Grace, M. G., Howe, G. R., and Miller, A. B.,** A case-control study of diet and colorectal cancer, *Int. J. Cancer*, 26, 757, 1980.
30. **Miller, A. B., Howe, G. R., Jain, M, Craib, K. J. P., and Harrison, L.,** Food items and food groups as risk factors in a case-control study of diet and colorectal cancer, *Int. J. Cancer*, 32, 155, 1983.
31. **Modan, B., Barrel, V., Lubin, F., Modan, M., Greenberg, R. A., and Graham, S.,** Low fiber intake as an etiologic factor in cancer of the colon, *J. Natl. Cancer Inst.*, 55, 15, 1975.
32. **Dales, L. G., Friedman, G. D., Wry, H. K., Grossman, S., and Williams, S. R.,** Case-control study of relationships of diet and other traits to colorectal cancer in American Blacks, *Am. J. Epidemiol.*, 109, 132, 1979.
33. **Reddy, B. S., Hedges, A. R., Laakso, K., and Wynder, E. L.,** Metabolic epidemiology of large bowel cancer: fecal bulk and constituents of high-risk North American and low-risk Finnish population, *Cancer*, 42, 2832, 1978.
34. **Domellof, L., Darby, L., Hanson, D., Simi, B., and Reddy, B. S.,** Fecal sterols and bacterial β-glucuronidase activity: a preliminary metabolic epidemiology study of healthy volunteers from Umea, Sweden, and metropolitan New York, *Nutr. Cancer*, 4, 120, 1982.
35. **Reddy, B. S., Ekelund, G., Bohe, M., Engle, A., and Domellof, L.,** Metabolic epidemiology of colon cancer: dietary pattern and fecal sterol concentrations of three populations, *Nutr. Cancer*, 5, 34, 1983.
36. **Bingham, S., Williams, D. R. R., Cole, T. J., and James, W. P. T.,** Dietary fiber and regional large-bowel cancer mortality in Britain, *Br. J. Cancer*, 40, 456, 1976.
37. **Reddy, B. S.,** Tumor promotion in colon carcinogenesis, in *Mechanisms of Tumor Promotion*, Vol. 1, Slaga, T., Ed., CRC Press, Boca Raton, Fla., 1983, 107.
38. **Hill, M. J., Drasar, B. S., Aries, V. C., Crowther, J. S., Hawksworth, G. B., and Williams, R. E. O.,** Bacteria and etiology of cancer of large bowel, *Lancet*, 1, 95, 1971.
39. **Aries, V., Crowther, J. S., Drasar, B. S., Hill, M. J., and Williams, R. E. O.,** Bacteria and etiology of cancer of the large bowel, *Gut*, 10, 334, 1969.
40. **Reddy, B. S.,** Dietary lipids and their relationship to colon cancer, in *Diet, Nutrition, and Cancer: From Basic Research to Policy Implications*, Roe, D. A., Ed., Alan R. Liss, New York, 1983.
41. **Wattenberg, L. W.,** Inhibitors of neoplasia by minor dietary constituents, *Cancer Res.*, 43, 2448 S, 1983.
42. **Ehrich, M., Ashell, J. E., Van Tassell, R. L., Wilkins, T. D., Walker, A. R. P., and Richardson, N. J.,** Mutagens in the feces of 3 South African populations at different levels of risk for colon cancer, *Mutat. Res.*, 64, 231, 1979.
43. **Bruce, W. R., Varghese, A. J., Furrer, R., and Land, P. C.,** A mutagen in the feces of normal humans, in *Origins of Human Cancer*, Vol. 4, Cold Spring Harbor Conf. Cell Proliferation, Hiatt, H. H., Watson, J. P., and Winston, J. A., Eds., Cold Spring Harbor Laboratory, Cold Spring Harbor, New York, 1977, 1641.
44. **Reddy, B. S., Sharma, C., Darby, L., Laakso, K., and Wynder, E. L.,** Metabolic epidemiology of large bowel cancer: fecal mutagens in high- and low-risk population for colon cancer, a preliminary report, *Mutat. Res.*, 72, 511, 1980.

45. **Reddy, B. S., Sharma, C., Mathews, L., and Engle, A.,** Fecal mutagens from subjects consuming a mixed-western diet, *Mutat. Res.,* 135, 11, 1984.
46. **Mower, H. F., Ichinotsubo, D., Wang, L. W., Mandel, M., Stemmerman, A., Nomura, A., Heilbrun, L., Kamiyama, S., and Shimada, A.,** Fecal mutagens in two Japanese populations with different colon cancer risks, *Cancer Res.,* 42, 1164, 1982.
47. **Kuhnlein, U., Bergstrom, D., and Kuhnlein, H.,** Mutations in feces from vegetarians and non-vegetarians, *Mutat. Res.,* 45, 1, 1981.
48. **Hirai, N. and Kingston, D. G. I.,** Structure elucidation of a potent mutagen from human feces, *J. Am. Chem. Soc.,* 104, 6149, 1982.
49. **Fink, D. J. and Kritchevsky, D., Eds.,** Workshop on Fat and Cancer, *Cancer Res.,* 41, 3684, 1981.
50. **Reddy, B. S.,** Dietary fat and its relationship to large bowel cancer, *Cancer Res.,* 41, 3700, 1981.
51. **Vahouny, G. V. and Kritchevsky, D., Eds.,** *Dietary Fiber in Health and Disease,* Plenum Press, New York, 1982.
52. **Reddy, B. S.,** Dietary fiber and colon carcinogenesis: a critical review, in *Dietary Fiber in Health and Disease,* Vahouney, G. V. and Kritchevsky, D., Eds., Plenum Press, New York, 1982, 265.
53. **Reddy, B. S. and Wynder, E. L.,** Large bowel carcinogenesis: fecal constituents of populations with diverse incidence rates of colon cancer, *J. Natl. Cancer Inst.,* 50, 1437, 1973.
54. **Jensen, O. M., MacLennan, R., and Wahrendorf, J.,** Diet, bowel function, fecal characteristics and large bowel cancer in Denmark and Finland, *Nutr. Cancer,* 4, 5, 1982.
55. **Weisburger, J. H. and Fiala, E. S.,** Experimental colon carcinogenesis and their mode of action, in *Experimental Colon Carcinogenesis,* Autrup, H. and Williams, G. M., Eds., CRC Press, Boca Raton, Fla., 1983, 27.
56. **Shamsuddin, A. K.,** In vivo induction of colon cancer dose and animal species, in *Experimental Colon Carcinogenesis,* Autrup, H. and Williams, G. M., Eds., CRC Press, Boca Raton, Fla., 1983, 51.
57. **Reddy, B. S., Watanabe, K., and Weisburger, J. H.,** Effect of high-fat diet on colon carcinogenesis in F344 rats treated with 1,2-dimethylhydrazine, methylazoxymethanol acetate or methylnitrosourea, *Cancer Res.,* 37, 416, 1977.
58. **Reddy, B. S. and Ohmori, T.,** Effect of intestinal microflora and dietary fat on 3,2'-dimethyl-4-amino-biphenyl-induced colon carcinogenesis in F344 rats, *Cancer Res.,* 41, 1363, 1981.
59. **Bull, A. W., Soullier, B. K., Wilson, P. S., Haydon, M. T., and Nigro, N. D.,** Promotion of azoxymethane-induced intestinal cancer by high-fat diets in rats, *Cancer Res.,* 39, 4956, 1979.
60. **Wargovich, M. J. and Felkner, I. C.,** Metabolic activiation of DHM by colonic microsomes: a process influenced by type of fat, *Nutr. Cancer,* 4, 146, 1982.
61. **Nigro, N. D., Singh, D. V., Campbell, R. L., and Pak, M. S.,** Effect of dietary beef fat on intestinal tumor formation by azoxymethane in rats, *J. Natl. Cancer Inst.,* 54, 429, 1975.
62. **Bansal, B. R., Rhoads, J. E., Jr., and Bansal, S. C.,** Effects of diet on colon carcinogenesis and the immune system in rats treated with 1,2-dimethylhyrazine, *Cancer Res.,* 38, 3293, 1978.
63. **Rogers, A. E. and Newberne, P. M.,** Dietary effects of chemical carcinogenesis in animal models for colon and liver tumors, *Cancer Res.,* 35, 3427, 1975.
64. **Reddy, B. S. and Maeura, Y.,** Tumor promotion by dietary fat in azoxymethane-induced colon carcinogenesis in female F344 rats: influence of amount and source of dietary fat, *J. Natl. Cancer Inst.,* 72, 745, 1984.
65. **Nauss, K. M., Locniskoor, M., and Newberne, P. M.,** Effect of alterations in quality and quantity of dietary fat on 1,2-dimethylhydrazine-induced colon tumorigenesis in rats, *Cancer Res.,* 43, 4083, 1983.
66. **Watanabe, K., Reddy, B. S., Wong, C. Q., and Weisburger, J. H.,** Effect of undergraded carrageenan on colon carcinogenesis in F344 rats treated with azoxymethane or methylnitrosourea, *Cancer Res.,* 38, 4427, 1978.
67. **Glauert, H. P., Bennink, M. R., and Sander, C. H.,** Enhancement of 1,2-dimethylhydrazine-induced colon carcinogenesis in mice by dietary agar, *Food Cosmet. Toxicol.,* 19, 281, 1981.
68. **Reddy, B. S., Narisawa, T., Vukusich, D., Weisburger, J. H., and Wynder, E. L.,** Effect of quality and quantity of dietary fat and dimethylhydrazine in colon carcinogenesis in rats, *Proc. Soc. Exp. Biol. Med.,* 151, 237, 1976.
69. **Broitman, S. A., Vitale, J. J., Vavrousek-Jakuba, E., and Gottlieb, L. S.,** Polyunsaturated fat, cholesterol and large bowel tumorigenesis, *Cancer,* 40, 2455, 1977.
70. **Sakaguchi, M., Hiramatsu, Y., Takada, H., Yamamura, M., Hioki, K., Saito, K., and Yamamoto, M.,** Effect of dietary unsaturated and saturated fats on azoxymethane-induced colon carcinogenesis in rats, *Cancer Res.,* 44, 1472, 1984.
71. **Cruse, J. P., Lewin, M. R., and Clark, C. G.,** Failure of bran to protect against experimental colon cancer in rats, *Lancet,* 2, 1278, 1978.
72. **Wilson, R. B., Hutcheson, D. P., and Wideman, L.,** Dimethylhydrazine-induced colon tumors in rats fed diets containing beef fat or corn oil with and without wheat bran, *Am. J. Clin. Nutr.,* 30, 176, 1977.

73. **Nigro, N. D., Bull, A. W., Klopfer, B. A., Pak, M. S., and Campbell, R. L.,** Effect of dietary fiber on azoxymethane-induced intestinal carcinogenesis in rats, *J. Natl. Cancer Inst.*, 62, 1097, 1979.

74. **Bauer, H. G., Asp, N., Oste, R., Dahlquist, A., and Fredlund, P.,** Effect of dietary fiber on the induction of colorectal tumors and fecal β-glucuronidase activity in the rat, *Cancer Res.*, 39, 3752, 1979.

75. **Jacobs, L. R.,** Enhancement of rat colon carcinogenesis by wheat bran consumption during the stage of 1,2-dimethylhydrazine administration, *Cancer Res.*, 43, 4057, 1983.

76. **Watanabe, K., Reddy, B. S., Weisburger, J. H., and Kritchevsky, D.,** Effect of dietary alfalfa, pectin, and wheat bran on azoxymethane or methylnitrosourea-induced colon carcinogenesis in F344 rats, *J. Natl. Cancer Inst.*, 63, 141, 1979.

77. **Reddy, B. S., Mori, H., and Nicolais, M.,** Effect of dietary wheat bran and dehydrated citrus fiber on azoxymethane-induced intestinal carcinogenesis in Fischer 344 rats, *J. Natl. Cancer Inst.*, 66, 553, 1981.

78. **Reddy, B. S. and Mori, H.,** Effect of dietary wheat bran and dehydrated citrus fiber on 3,2-dimethyl-4-aminobiphenyl-induced intestinal carcinogenesis in F344 rats, *Carcinogenesis*, 2, 21, 1981.

79. **Barnes, D. S., Clapp, N. K., Scott, D. A., Oberst, D. L., and Berry, S. G.,** Effects of wheat, rice, corn and soybean bran on 1,2-diemthylhydrazine-induced large bowel tumorigenesis in F344 rats, *Nutr. Cancer*, 5, 1, 1983.

80. **Reddy, B. S., Mauera, Y., and Wayman, M.,** Effect of dietary corn bran and autohydrolyzed lignin on 3,2'-dimethyl-4-aminobiphenyl-induced intestinal caracinogenesis in male F344 rats, *J. Natl. Cancer Inst.*, 71, 419, 1983.

81. **Freeman, H. J.,** Dietary fibers and colon cancer, in *Experimental Colon Carcinogenesis*, Autrup, H. and Williams, G. M., Eds., CRC Press, Boca Raton, Fla., 1983, 267.

82. **Reddy, B. S., Mangat, S., Sheinfil, A., Weisburger, J. H., and Wynder, E. L.,** Effect of type and amount of dietary fat and 1,2-dimethylhydrazine on biliary bile acids, fecal bile acids and neutral sterols in rats, *Cancer Res.*, 37, 2132, 1977.

83. **Nigro, N. D. and Bull, A. W.,** The two-step concept of intestinal carcinogenesis, in *Experimental Colon Carcinogenesis*, Autrup, H. and Williams, G. M., Eds., CRC Press, Boca Raton, Fla., 1983, 215.

84. **Reddy, B. S., Watanabe, K., and Sheinfil, A.,** Effect of dietary wheat bran, alfalfa, pectin and carrageenan on plasma cholesterol and fecal bile acid and neutral sterol excretion in rats *J. Nutr.*, 110, 1247, 1980.

85. **Narisawa, T., Magadia, N. E., Weisburger, J. H., and Wynder, E. L.,** Promoting effect of bile acid on colon carcinogenesis after intrarectal instillation of MNNG in rats, *J. Natl. Cancer Inst.*, 53, 1093, 1974.

86. **Reddy, B. S., Narisawa, T., Weisburger, J, H., and Wynder, E. L.,** Promoting effect of deoxycholic acid on colonic adenocarcinomas in germfree rats, *J. Natl. Cancer Inst.*, 56, 441, 1976.

87. **Reddy, B. S., Watanabe, K., Weisburger, J. H., and Wynder, E. L.,** Promoting effect of bile acids in colon carcinogenesis in germfree and conventional F344 rats, *Cancer Res.*, 37, 3238, 1977.

88. **Cohen, B. I., Raicht, R. F., Deschner, E. E., Takahashi, M., Sarwal, A. N., and Fazini, E.,** Effect of cholic acid feeding on N-methyl-N-nitrosourea-induced colon tumors and cell kinetics in rats, *J. Natl. Cancer Inst.*, 64, 573, 1980.

89. **Takano, S., Akagi, M., and Bryan, G. T.,** Stimulation of ornithine decarboxylase activity and DNA synthesis by phorbol esters or bile acids in rat colon, *Gann*, 75, 29, 1984.

90. **Lipkin, M.,** Cell proliferation in colon carcinogenesis, in *Experimental Colon Carcinogenesis*, Autrup, H. and Williams, G. M., Eds., CRC Press, Boca Raton, Fla. 1983, 139.

91. **Wynder, E. L.,** Reflections on diet, nutrition and cancer, *Cancer*, 43, 3024, 1983.

92. **Palmer, S.,** Diet, nutrition, and cancer: the future of dietary policy, *Cancer Res.*, 43, 2509 S, 1983.

93. **DeWys, W. D. and Greenwald, P.,** Clinical trials: a recent emphasis in the prevention program of the National Cancer Institute, *Semin. Oncol.*, 10, 300, 1983.

94. **Paul, A. A. and Southgate, D. A. T., Eds.,** McCance and Widdowson's The Composition of Foods, Her Majesty's Stationery Office, London, 1978.

95. **Judd, J. P., Kelsay, J. L., and Mertz, W.,** Potential risks from low-fat diets, *Semin. Oncol.*, 10, 273, 1983.

Chapter 5

NUTRITION AND THE EPIDEMIOLOGY OF BREAST CANCER

A.B. Miller

Breast cancer is the most important cause of death from cancer in women in the Western World, though in some countries, this position will shortly be overtaken by lung cancer. Of the 104 sets of incidence data from different registries and racial groups within registries published in Volume IV of Cancer Incidence in Five Continents,[1] breast cancer shows the highest incidence in women in 94. Breast cancer is well recognized as hormonally associated[2-4] but diet for some time has been suspected as an important cause.[2,5]

Several lines of evidence support the importance of nutritional and dietary factors in the etiology of breast cancer. The first derives from animal experimental studies, which have demonstrated that in the presence as well as the absence of mammary carcinogens, the incidence of mammary tumors in rats increases substantially with high-fat diets, providing the diet contains a small amount of unsaturated fat. This evidence is reviewed further in Chapter 6.

The second type of evidence comes from studies of changing incidence and mortality (for example in Iceland[6] and Japan[7]) that suggest the effects of changing environmental factors in a country in increasing the incidence of breast cancer. The studies of migrants[8,9] particularly suggest the effect of acculturation, usually with a slow change in incidence over several generations. In Japanese migrants to California, the incidence of breast cancer in premenopausal women has now almost reached that of the Caucasian population.[10] Many aspects of lifestyle change with the acculturation to the host country which follows migration. However, nutritional aspects of lifestyle are one of the most important, especially if the migrants are rapidly absorbed into the surrounding cultural milieu. This may be happening in Israel, where substantial changes are occurring in the rates of cancer in groups that have migrated there from Africa and Asia, together with a rise in the incidence in the Israeli born up to the levels experienced by those born in Europe and America.[10] These changes are occurring almost as rapidly for breast cancer as they are for colorectal cancer, suggesting that some influences on breast cancer do not necessarily operate only in early life or adolescence, as had tended to be assumed from the changes following migration of the Japanese. The studies of special religious groups[11,12] also suggest the influence of cultural or lifestyle and particularly dietary factors on the etiology of breast cancer. Interestingly, cohort studies of religious groups who entered orders or changed to a special group in adult life have not shown lower breast cancer rates than expected from the general population. This is true for Adventist women in California[13] as well as members of strict religious orders in Britain who eschew or eat very little meat.[14] This suggests that dietary factors in these groups operated before adult life, or alternatively that the use of high-fat dairy products in these groups in substitution for meat maintained a total fat intake little less than normal. Kinlen[14] attempted to evaluate the latter in his study of nuns by making a crude estimate of dietary fat intake. He found no evidence of an association. Kinlen et al.[15] also failed to find any indication of lower proportions of breast cancer deaths, or for that matter deaths from cardiovascular disease, among members of a vegetarian society. Nevertheless, Moolgavkar et al.[16] have shown that changes in incidence of breast cancer in different populations can be related to changes in successive birth cohorts, supporting the possibility that patterns established in relatively early life are critical.

The third line of evidence comes from population correlation studies associating breast cancer incidence and mortality with total dietary fat and other nutrient intake. It is important that not only have such correlations been noted internationally[17-20] but also within Japan[8] as well as the U.S.[21] for total fat even though the latter report failed to find the correlation with animal fat that had been found in the international studies. Hirayama[7] commented that dietary fat intake appears to have shown the most striking increase of all the nutritional changes that have been noted in Japan in recent years. Gaskill et al.[22] found a positive correlation between breast cancer mortality and milk, table fats, beef, calories, protein, and fat and a negative correlation with egg consumption within the U.S. Only the association of milk and egg consumption remained significant, however, when age of first marriage (as an indicator of age at first birth) was controlled. These two associations also persisted after controlling for other demographic and dietary variables including intake of fat. This study thus suggested a special role for dairy products in the etiology of breast cancer. Hems[23] found age-standarized breast-cancer mortality rates for women of 41 countries during 1970 to 1971 were positively correlated with total fat, animal protein, and animal calories, independently of other components of diet for 1964 to 1966. Differences in childbearing appeared to contribute little to the variation of breast cancer mortality rates between countries. Hems[24] subsequently evaluated changes of breast cancer mortality for women in England and Wales between 1911 and 1975 in relation to changes in the consumption of fat, sugar, and animal protein 1 to 2 decades earlier. The association was strongest for fat and sugar intake 1 decade earlier. The mortality changes were not related to changes in childbearing. He noted that the social class gradient in breast cancer mortality almost disappeared during the 1950's, rates declining for the upper classes but increasing for the lower. These changes could have resulted from the changes that occurred in the diet of the different classes in the early 1940's.

Schrauzer[25,26] postulated an inverse relationship between mortality from breast and colon cancer and the consumption of cereals and seafoods, according to food consumption tables published by the Food and Agricultural Organization. There appeared to be a direct correlation between these cancers and a high intake of fat, sugar, and meat.[27] Schrauzer suggested that food consumption patterns reflected the selenium content of the diets consumed and postulated a negative correlation between selenium intake of population groups and their mortality from cancer. This hypothesis was later modified to take into account a postulated antagonistic effect of dietary zinc on selenium, and it was predicted that a twofold increase of selenium intake in the U.S. would reduce breast cancer to about 10% of its present incidence. However, the validity of calculating selenium intake from food consumption tables and assigning an arbitrary, average, selenium value to the major foods may be questioned. The selenium concentration in grains, for example, can vary more than tenfold, depending on the environment in which the crop was grown.

The fourth, indirect line of evidence comes from the effect of certain risk factors for breast cancer, which are particularly probably nutritionally mediated. These are particularly weight, height (and the related indices of body mass dependent on height and weight), and age at menarche.

As a result of a case-control study of 632 white Americans and 1253 hospital controls, Wynder et al.[28] suggested that stocky, in particular somewhat obese women, have a slightly higher chance of developing breast cancer than those of slimmer body build. However Wynder[29] later pointed out that women in a low socioeconomic class have three to four times more obesity than women in a high socioeconomic class. After matching for socioeconomic factors, he found no clear association between obesity and breast cancer.

Brinkley et al.[30] attempted an anthropometric evaluation of the possibility that breast cancer is associated with a particular body type. They included 150 patients with breast cancer, and three comparison groups: women with other cancers, other diseases, and normal women. Weight, height, sitting height, biacromial dimension, and biiliac dimension were measured. Various indices of these variables were calculated, a multivariate analysis conducted, and a discriminant function estimated. Weight, sitting height, and biacromial-biiliac ratio were among the variables significant in the analysis which discriminated the cases from the controls, and it was concluded that women with breast cancer tend to be more masculine in type.

In some of the centers in an International collaborative case-control study, an association was found between height, weight, and breast cancer. In Athens[31] 799 cases and 2470 hospital controls were included. Increasing risk was found with increasing height and weight, and with the index weight/height.[2] Adjustment for variables associated with height and weight (including schooling as an index of a socioeconomic status) failed to abolish this effect. The authors concluded that both height and weight have independent associations with breast cancer risk, the tallest and heaviest deciles of the population having almost twice the risk experienced by the shortest and lightest deciles.

In Sao Paulo,[32] where 536 cases of breast cancer and 1550 hospital controls were interviewed, the data were subdivided by age. An association of increasing risk with increasing weight was found in women age 50 or more only, and not in women age 20 to 49. This association was reduced, but not abolished when the index weight/height[2] was used to control for height. In a parallel study of 213 cases of breast cancer and 648 hospital controls in Taiwan[33] there was a significant association of increasing risk with increasing weight, but not for height. Women weighing more than 55 kg had more than twice the breast cancer risk of women weighing less than 45 kg. When the data were subdivided by age, there was a suggestion of a stronger effect of weight in woman older than 50, but the effect in the younger women was not abolished.

The hypothesis that breast cancer in postmenopausal women was associated with obesity through a hormonal mechanism was proposed by de Waard and colleagues,[34] based on cytologic studies of the vaginal smears of obese postmenopausal women that suggested continuing estrogenic activity. They studied 108 patients with breast cancer, and 571 women from the general population. Obesity was defined as body weight more than 25% above ideal weight. In women age 55 or more 71% of those with breast cancer were obese compared to 54% of the controls, a significant difference.

deWaard subsequently refined his hypothesis to postulate two types of breast cancer.[35] He suggested that one, occurring largely in premenopausal women, is connected with an endocrine imbalance in which the ovarian hormones are involved, the other, largely occurring in postmenopausal women, has as its major determinant altered hormonal homeostasis related to overnutrition. This hypothesis was supported by a prospective study of 7259 postmenopausal women followed on average for 5.4 years, in whom 70 cases of breast cancer occurred.[36] Increasing risk with increasing weight and increasing height was found, with women weighing 70 kg or more and 165 cm or more in height having 3.6 times the risk of women weighing less than 60 kg and less than 160 cm in height. However, using Quetelets index (weight/height[2]) as a measure of degree of overweight as a correction for height, removed a large degree of the risk. Therefore deWaard[37] modified his hypothesis to postulate that body mass rather than overweight (obesity) was the risk factor.

In a further study[38] deWaard et al. evaluated the effect of weight and height on the age-specific incidence of breast cancer using data from the Netherlands and Japan. They computed age-specific curves for different height and weight groups. There was a divergence of the curves in postmenopausal women with the heavier and taller woman showing higher incidence levels. There appeared to be little independent effect for weight if there was an adjustment for its correlation with height. deWaard[37] suggested that lean body mass may be the important variable. However if height is critical (and it is critical to the calculation of lean body mass) the relevant nutritional factors must begin to operate in adolescence or earlier, rather than in adult life, as was pointed out by MacMahon.[39] deWaard et al.[38] suggested that about one half of the differences in incidence between Holland and Japan could be attributed to differences in body weight and height.

In a case control study of 400 cases of breast cancer and 400 neighborhood controls in Canada, hardly any effect of height, but an effect of weight in postmenopausal women when assessed (by questionnaire recall) 12 months before diagnosis and at the time of the menopause was found.[40] The largest difference was in a small group of cases and controls age 70 or more. There were no differences between premenopausal cases and controls. Later we attempted to replicate these findings in women with breast cancer age 65 or more and controls drawn from the population, but only found a small (and nonsignificant) weight differential, with no greater effect in women age 70 or more.[41]

Support for the hypothesis of deWaard has come from a study of 1868 breast cancer patients and 3391 controls.[42] In premenopausal women, lowered risk of breast cancer with higher Quetelet's index was found. In contrast, in postmenopausal women a markedly higher risk of breast cancer associated with increased Quetelet's index was found. Further, although an estimate of weight gain between age 20 and interview showed little or no influence on breast cancer risk in the premenopausal woman, it was strongly related to risk in postmenopausal women.

Further, in the long-term prospective study conducted by the American Cancer Society, a significant trend of increasing mortality from breast cancer with increasing weight index was found.[43] For those with a weight index of 140+ there was a mortality ratio of 1.53 in relation to those 90 to 109% of average weight, in contrast to a mortality ratio of 0.82 for those with a weight index of less than 80 (the severely underweight). In addition, in U.S. breast cancer screening program participants, 405 patients with breast cancer identified in the first 2 years were compared to 1156 normal screenees.[44] There was increasing risk with increasing weight, but little effect of height when adjusted for weight.

Completely negative findings have also been reported for height and weight. Thus, in a case control study of 179 newly diagnosed patients with breast cancer and age matched controls from a population register in Sweden, no difference in the distributions of height and weight or two different indices for overweight was found.[45] Further, in a large case control study using other cancer cases as controls for cases with breast cancer, Wynder et al.[46] failed to find an effect of weight in woman with pre-, peri-, or postmenopausal breast cancer. Soini[47] in a case control study of 122 cases of breast cancer and 534 controls age 41 to 60 from a breast cancer screening program found no effect of height, weight, degree of overweight (Quetelet's index), or the product weight × height. However as only 27 of the cases of breast cancer came from the screening program (the remainder were referrals to the radiotherapy department) the comparability of the cases and controls must be in doubt and a possible effect could have been matched out. Nevertheless, Hakama et al.[48] found no association of height, weight, the product height × weight, and body mass (weight per height^2) in a population study of females in selected municipalities to determine the correlation of these estimators of the size of the women with breast cancer incidence. They concluded that factors which are reflected by the standard of living and fertility might act independently and not through nutritional status, for which they regarded the size of the women as defined by the variables they measured, as an operational indicator. Nevertheless, seeking an effect of nutrition using indirect indicators at the population level is an insensitive approach and therefore this study can not be regarded as refuting the hypothesis.

Age at menarche is a risk factor for breast cancer, though the effect is relatively weak.[2,3] Women with an early age at menarche, especially prior to the age of 12, have the highest risk. There is evidence that body weight and food intake are related to early estrus of rats[49] supportive of the hypothesis that a critical body composition of fatness is essential for estrus in the rat, as it appears to be for age at menarche in the human female.[50] Hence the effect of diet and nutrition on breast cancer, if it is mediated through a hormonal mechanism[51] could be at least partly through age at menarche.[4]

Petrakis et al.[52] investigated nipple aspirates of breast fluid from nonpregnant healthy women. Cholesterol levels were found to be elevated above plasma levels and to increase with advancing age. Cholesterol epoxide, a carcinogen in animals, was detected in 7 of 17 women, most of whom had high levels of breast fluid cholesterol. Petrakis hypothesized that high-dietary fat may increase the level of cholesterol in breast fluid with local derivation of carcinogenic substances such as cholesterol epoxide.

One of the difficulties of many of the studies performed to date, is a failure of the investigators to consider the effect of multiple variables and their interrelationships. In the past this was often due to the complexity of the required analytic procedures so that most analyses were conducted in the univariate mode. In addition, many of the available data sets were either too small or contained too little information on many of the variables that should be considered, to permit such analyses. Gray et al.[53] however attempted in a population type correlation analysis to evaluate the effect of total fat and animal protein consumption on breast cancer incidence and mortality rates internationally while controlling for height, weight, and age at menarche. Although this was a study at the "macro" population level, they found that a significant effect of the dietary variables persisted after controlling for the other factors. This suggests that although some of the effects of diet on breast cancer may be mediated through effects on anthropometric and other risk factors such as age at menarche, there would seem to be more direct effects as well.

The final, and presumably most conclusive evidence, should come from studies of diet and nutrition at the individual level, either of the case-control or cohort type. Difficulties with dietary methodology have tended to inhibit investigations of this type. So far no cohort studies have been reported and only four case-control studies.

One case-control study involved 77 breast cancer cases and controls.[54] The cases were discharged from two Adventist operated hospitals and the controls, where possible, comprised two for each case selected from hospitalized cases of hernia and osteoarthritis and a third from the general Seventh-Day Adventist population. The food questions in the interview were designed to test the hypothesis that intake of high fat, low fiber, or both are associated with risk for breast cancer. Five foods were associated with breast cancer; fried foods, fried potatoes, hard fat frying, dairy products except milk, white bread, with relative risks ranging from 1.6 to 2.6. That for fried potatoes was highest and highly significant.

The second was a case-control study in four areas of Canada involving 400 newly diagnosed cases and 400 neighborhood controls.[55] Although three different approaches to assessment of diet were used (a 24-hr recall, a 4-day diary, and a detailed quantitative diet history) the final results were derived from the diet history, as this gave a higher response than the diary from the controls, and was directed to a period 6 months prior to the time of interview, and thus prior to the point of diagnosis of the cases. In a preliminary report,[5] results from a mean of the diary and dietary history were given, which suggested a stronger association with fat intake than in the final report.[55] The mean nutrient intake as estimated by the dietary history alone for six nutrients (total calories, total fat, saturated fat, oleic acid, linoleic acid, and cholesterol) was greater for the cases as a group than the controls though the differences were not statistically significant when the pre- and post menopausal groups were considered independently. In a risk ratio analysis, the strongest association in the premenopausal group was for total fat with weaker associations for saturated fat and cholesterol. When the effect of each nutrient was controlled for the effect of the others, the association for total fat increased in strength while those for saturated fat and cholesterol diminished. In the postmenopausal group, the only consistent finding was for total fat intake.

The risk ratios were low (1.6 for total fat in premenopausal women and 1.5 for postmenopausal women) and there was no evidence of a dose-response relationship. However, using these data and a weighted average over pre- and postmenopausal women, Miller[3] subsequently reported an attributable risk of 27% for total fat intake.

In a re-analysis of these data, Howe[56] has found increasing risk with increasing consumption of saturated fat, significant at the $p = 0.02$ level. This analysis followed the suggestion of Marshall and Graham[57] that increased precision would follow from calculating means of two or more different relatively imprecise methods. Howe[56] confirmed that combining the data from the 24-hr recall and diet history (6 months before interview) resulted in such increased precision. Interestingly, the strength of the association and the significance of the trend seemed greater in premenopausal than in postmenopausal women.

Graham et al.[58] reported the analysis of food frequency questionnaires administered to 2024 cases of breast cancer and 1463 hospital controls without cancer at Roswell Park Memorial Institute from 1958 to 1965. No association of breast cancer with estimated consumption of animal fat or other dietary factors was found. However this was a relatively brief questionnaire with estimates of fat consumption based on standard portion sizes and was not specifically designed to assess the fat hypothesis.

Lubin et al.[59] reported the results of a case-control study in northern Alberta involving 577 women age 30 to 80 with breast cancer interviewed in 1976 to 1977, and 826 disease-free controls interviewed subsequently. The questionnaire included information on the frequency with which eight food items and milk and butter were usually consumed. The major sources of animal fat and animal protein were represented. Using standard portion sizes estimates of consumption of animal fat were computed. Significant increasing trends of risk with more frequent consumption of beef, pork, and sweet desserts were found. Elevated risks were also noted for the use of butter at the table and for frying with butter and margarine, as opposed to vegetable oils. Risk also increased significantly with increasing indices of consumption of animal fat and animal protein, though the trends were not uniform, but not with cholesterol intake.

In an indirect attempt to assess the diet of women with breast cancer, Nomura et al.[60] compared the diet of 86 Japanese men whose wives had developed breast cancer with that of the remaining 6774 men in Hawaii who had had dietary data collected in the Japan-Hawaii Cancer Study. They assumed that there is a similarity between the diet of husbands and wives. They found that the husbands of the cases consumed more beef or meat, butter/margarine/cheese, corn, and weiners and less Japanese foods than the control spouses.

Currently two case-control studies of diet and breast cancer supported by the U.S. breast cancer task force (in Israel and Hawaii) are ongoing and a further case-control study in Recife, Brazil,[70] where there is a surprisingly high incidence of breast cancer. Only preliminary findings from the case-control study in Hawaii have so far been reported.[61] However a positive association between breast cancer risk and the intake of dietary fat (particularly saturated fat) and animal protein in postmenopausal women, especially the Japanese, was noted.

One possible mechanism for the effect of high-fat diets increasing the risk of breast cancer would be through changes in serum cholesterol and/or plasma lipids. However, prospective studies of women who have had such measurements have failed to find evidence of an association.[62]

Nevertheless, if the relevant mechanism is dependent on the levels of fat in the intestine (see Cohen, next chapter) differences in cholesterol or blood lipids are not only likely to be minimal but irrelevant. Even so, a study of the incidence and mortality of cancer of the breast among women who had previously had total colectomy and terminal ileostomy, mostly for ulcerative colitis, or Crohn's disease, failed to find a lower incidence of breast cancer, though there was a slight deficiency of deaths from this cause.[63] These findings suggest that, at least in adult life, the colon may not be involved in the etiology of breast cancer.

In conclusion, a number of different sources of information now support the association of diet, especially high fat in the diet, with risk of breast cancer. Although further work is desirable to clarify the association, the increasing consistency of the evidence derived from the epidemiologic studies and the similar data derived from experimental studies, suggest that the association is causal. Indeed, this evidence, together with similar evidence relating to colorectal and other cancers, led a committee of the U.S. National Academy of Sciences[64] to recommend attempts to achieve population reduction of dietary fat intake from the present level of 40 to 30% of available calories. This reduction seems achievable without a major disturbance in dietary habits, is unlikely to be hazardous, and after a period, should result in a reduction in the incidence of breast and other fat-associated cancers. It is important that early action be taken to ensure that the population makes appropriate changes in their diet. If, as some of the migrant and other descriptive studies suggest, much of the effect of diet and nutrition in increasing the risk of breast cancer operates early in life, there could be a substantial delay before the full impact of primary prevention is seen in the population.[65]

However, clearly more discriminating data on the desirable level of fat intake that should be regarded as the optimum for human populations is needed. We need to resolve in man the relative importance of different types of fat, particularly the possible opposed effects of unsaturated and saturated fats. We also need to determine the dose response for cancer induction for total, unsaturated, and saturated fats. Thus we need to answer the question "At what level does further protection from (breast) cancer from lowering fat intake cease?" Does it cease at 30% as recently recommended by the NAS committee,[64] or should we aim for 25% or even 20% or less as advocated by others? We also need to be concerned over the possibility that adverse effects on health, either through increased incidence of other cancer sites, unexpected effects on cardiovascular disease, etc. might offset the benefits derived by reducing the incidence of fat-associated cancers.

Where is the evidence required likely to come from? The NAS Committee on Diet Nutrition and Cancer have published their recommendations for further research in this area.[66] Further observational analytic epidemiologic studies are clearly needed, both of the case-control and cohort type. Case-control studies could profitably be performed in different populations with differing diets and breast cancer incidence. Cohort studies in which dietary data are collected well in advance of the diagnosis offer the advantage that the dietary information may well be collected at a more relevant time period to the natural history of the cancer, but also suffer from the disadvantages of less complete dietary information, expense, and long duration of follow-up required.

Carefully planned intervention studies of dietary factors that may prevent cancer induction should now be initiated in man. For breast cancer and fat, this will require detailed instruction in the ways to achieve a low-fat diet. Although far from simple, some groups have already initiated pilot studies in this area. Wynder and Cohen[67] have recommended an intervention study of fat reduction in women with stage 2 breast cancer, as there is some experimental evidence that supports such an approach in experimental animals (see Chapter 6). However, a follow-up of 300 of the cases in our case-control study of diet and breast cancer has failed to find any influence of fat on prognosis of breast cancer, though weight, and possibly particularly obesity, was associated with prognosis.[68] This supports a suggestion by de-Waard[69] that weight reduction should be assessed as a means of improving prognosis in women with breast cancer. If intervention studies of reduction in fat intake in women at high risk for breast cancer do proceed they will have to be carefully planned. They should involve randomization of subjects to experimental and control groups whenever possible. Care will have to be taken to define high risk in a way relevant to the hypothesis under test. Thus it does not seem sensible to define this by, for example, family history. Rather, some attempt should be made to select women at risk by virtue of a high-fat intake. The intervention should also be carefully planned to avoid dilution. Thus care will have to be taken to ensure compliance with the intervention in the study group but lack of intervention in the control group. To do this will not be easy, especially if informed consent of all subjects is insisted upon. Further, the intervention will not have to be so complex that it would be impossible to use this as an approach to control of breast cancer in the general population subsequently.

This is a formidable list of problems, but care taken in the design stage of a trial will bring dividends later. Some will object to the idea of any trial, suggesting rather that we should act now to advise the public on the actions they should take and monitor the subsequent impact of changes in diet on the cancer rates in the population. Probably both approaches can be supported, yet as the need now is for firmer evidence to justify major shifts of diet in the population, the intervention trial approach would seem to be the most scientifically justifiable.

REFERENCES

1. **Waterhouse, J., Muir, C., Shanmugaratnam, K., and Powell, J., Eds.,** Cancer Incidence in Five Continents, Vol. 4, IARC Scientific Publication No. 42, International Agency for Research on Cancer, Lyon, 1982.
2. **MacMahon, B., Cole, P., and Brown, J.,** Etiology of human breast cancer: a review, *J. Natl. Cancer Inst.*, 50, 21, 1973.
3. **Miller, A. B.,** An overview of hormone associated cancers, *Cancer Res.*, 38, 3985, 1978.
4. **Miller, A. B. and Bulbrook, R. D.,** Special report: the epidemiology and etiology of breast cancer, *N. Engl. J. Med.*, 303, 1246, 1980.
5. **Miller, A. B.,** Role of nutrition in the etiology of breast cancer, *Cancer*, 39, 2704, 1970.
6. **Bjarnson, O., Day, N., Snaedal, G., and Tulinius, H.,** The effect of year of birth on the breast cancer age incidence curve in Iceland, *Int. J. Cancer*, 13, 689, 1974.
7. **Hirayama, T.,** Changing patterns of cancer in Japan with special reference to the decrease of stomach cancer mortality, in *Origns of Human Cancer*, Cold Spring Harbor Laboratory, Cold Spring Harbor, N.Y., 1977, 55.
8. **Buell, P.,** Changing incidence of breast cancer in Japanese-American women, *J. Natl. Cancer Inst.*, 51, 1479, 1974.
9. **Dunn, J. E.,** Breast cancer amoung American Japanese in the San Francisco Bay area, *Natl. Cancer Inst. Monog.*, 47, 157, 1977.
10. **Steinitz, R.,** Cancer risks in immigrant population in Israel, in *Proceedings of the first UICC Conference on Cancer Prevention in Developing Countries*, Aoki, K., Ed., University of Nagoya Press, 1982, 363,.
11. **Lyon, J. L., Gardner, J. W., and West, D. W.,** Cancer risk and life-style: cancer among Mormons from 1967—1975, in *Cancer Incidence in defined Populations*, Banbury Rep. 4, Cairns, J., Lyon, J. L., and Skolnick, M., Eds., Cold Spring Harbor Laboratory, Cold Spring Harbor, N.Y., 1980, 3.
12. **Phillips, R. L., Kuzqma, J. W., and Lotz, T. M.,** Cancer mortality among comparable members versus non members of the Seventh Day Adventist Church, in *Cancer Incidence in Defined Populations*, Banbury Report 4, Cairns, J., Lyon, J. L., and Skolnick, M., Eds., Cold Spring Harbor Laboratory, Cold Spring Harbor, N.Y., 1980, 93.
13. **Phillips, R. L., Garfinkel, L., Kuzqma J. W., Beeson W. L., Lotz, T., and Brin, B.,** Mortality among California Seventh-Day Adventists for selected cancer sites, *J. Natl. Cancer Inst.*, 65, 1097, 1980.
14. **Kinlen, L. J.,** Meat and fat consumption and cancer mortality: a study of strict religious orders in Britain, *Lancet*, 1, 946, 1982.
15. **Kinlen, L. J., Herman, C., and Smith, P. G.,** A proportionate study of cancer mortality amoung members of a vegetarian Society, *Br. J. Cancer*, 48, 355, 1983.
16. **Moolgavkar, S. H., Day, N. E., and Stevens, R. G.,** Two-stage model for carcinogenesis: epidemiology of breast cancer in females, *J. Nat. Cancer Inst.*, 65, 559, 1980.
17. **Drasar, B. S. and Irving, D.,** Environmental factors and cancer of the colon and breast, *Br. J. Cancer*, 27, 167, 1973.
18. **Armstrong, B. and Doll, R.,** Environmental factors and cancer incidence and mortality in different countries, with special reference to dietary practices, *Int. J. Cancer*, 15, 617, 1975.
19. **Carroll, K. K.,** Experimental evidence of dietary factors and hormone-dependent cancers, *Cancer Res.*, 35, 3374, 1975.

20. **Knox, E. G.,** Foods and diseases, *Br. J. Soc. Prev. Med.,* 31, 71, 1977.
21. **Enig, M. G., Munn, R. J., and Keeney, M.,** Dietary fats and cancer trends — a critique, *Fed. Proc.,* 37, 2215, 1978.
22. **Gaskill, S. P., McGuire, W. L., Osborne, C. K., and Stern, M. P.,** Breast cancer mortality and diet in the United States, *Cancer Res.,* 39, 3628, 1979.
23. **Hems, G.,** The contributions of diet and child bearing to breast cancer rates, *Br. J. Cancer,* 37, 974, 1978.
24. **Hems, G.,** Associations between breast cancer mortality rates, child-bearing and diet in the United Kingdom, *Br. J. Cancer,* 41, 429, 1980.
25. **Schrauzer, G. N.,** Cancer mortality correlation studies. II. Regional associations of mortalities with the consumption of foods and other commodities, *Med. Hypoth.,* 2, 39, 1976.
26. **Schrauzer, G. N., White, D. A., and Schneider, C. J.,** Cancer mortality correlation studies. III. Statistical associations with dietary selenium intakes, *Bioinorg. Chem.,* 7, 23, 1977.
27. **Schrauzer, G. N.,** Selenium and cancer. A review, *Bioinorg. Chem.,* 5, 275, 1975.
28. **Wynder, E. L., Bross, I. J., and Hirayama, T.,** A study of the epidemiology of cancer of the breast, *Cancer,* 13, 559, 1960.
29. **Wynder, E. L.,** Identification of women at high risk for breast cancer, *Cancer,* 24, 1235, 1969.
30. **Brinkley, D., Carpenter, R. G., and Haybittle, J. L.,** An anthropometric study of women with cancer, *Br. J. Prev. Soc. Med.,* 25, 65, 1971.
31. **Valoras, V. G., MacMahon, B., Trichopoulos, D., and Polychronopoulou, A.,** Lactation and reproductive histories of breast cancer patients in greater Athens, 1965—67, *Int. J. Cancer,* 4, 350, 1969.
32. **Mirra, A. P., Cole, P., and MacMahon, B.,** Breast cancer in an area of high parity, *Cancer Res.,* 31, 77, 1971.
33. **Lin, T. M., Chen, K. P., and MacMahon, B.,** Epidemiologic characteristics of cancer of the breast in Taiwan, *Cancer,* 27, 1497, 1971.
34. **deWaard, F., Laive, J. W. J., and Baanders-van Haliwijn, E. A.,** On the bimodal age distribution of mammary carcinoma, *Br. J. Cancer,* 114, 437, 1960.
35. **deWaard, F., Baanders-van Haliwijn, E. A., and Huizinga, J.,** The bimodal age distribution of patients with mammary cancer, *Cancer,* 17, 141, 1964.
36. **deWaard, F., Baanders-van Halewijn, E. A.,** A prospective study in general practice on breast cancer risk in post menopausal women, *Int. J. Cancer,* 14, 153, 1974.
37. **deWaard, F.,** Breast cancer incidence and nutritional status with reference to body weight and height, *Cancer Res.,* 35, 3351, 1975.
38. **deWaard, F., Cornelis, J. P., Aoki, K., and Yoshida, M.,** Breast cancer incidence according to weight and height in two cities of the Netherlands and Aichi prefecture Japan, *Cancer,* 40, 1269, 1977.
39. **MacMahon, B.,** Formal discussion of breast cancer incidence and nutritional status with particular reference to body weight and height, *Cancer Res.,* 35, 3357, 1975.
40. **Choi, N. W., Howe, G. R., Miller, A. B., Matthews, V., Morgan, R. W., Munan, L., Burch, J. D., Feather, J., Jain, M., and Kelly, A.,** An epidemiologic study of breast cancer, *Am. J. Epidemiol.,* 107, 510, 1978.
41. **Burch, J. D., Howe, G. R., and Miller, A. B.,** Breast cancer in relation to weight in woman aged 65 years and over, *Can. Med. Assoc. J.,* 124, 1326, 1981.
42. **Paffenbarger, R. S., Kampert, J. B., and Chang, H. G.,** Characteristics that predict breast cancer before and after the menopause, *Am. J. Epidemiol.,* 112, 258, 1980.
43. **Lew, E. A. and Garfinkel, L.,** Variations in mortality by weight among 750,000 men and women, *J. Chron. Dis.,* 32, 563, 1979.
44. **Brinton, L. A., Williams, R. R., Hoover, R. N., Stegens, N. L., Feinleib, M., and Fraumeni, J. F.,** Breast cancer risk factors among screening program participants, *J. Natl. Cancer Inst.,* 62, 37, 1979.
45. **Adami, H. O., Rimsten, A., Stenkvist, B., and Vegelius, J.,** Influence of height, weight and obesity on risk of breast cancer in an unselected Swedish population, *Br. J. Cancer,* 36, 787, 1977.
46. **Wynder, E. L., MacGornach, F. A., and Stellman, S. D.,** The epidemiology of breast cancer in 875 United States Caucasian women, *Cancer,* 41, 2345, 1978.
47. **Soini, I.,** Risk factors of breast cancer in Finland, *Int. J. Epidemiol.,* 6, 365, 1977.
48. **Hakama, M., Soini, I., Kuosma, E., Lethonen, M., and Aromaa, A.,** Breast cancer incidence: geographical correlations in Finland, *Int. J. Epidemiol.,* 8, 33, 1979.
49. **Frisch, R. E., Hegsted, D. M., and Yoshinaga, K.,** Body weight and food intake at early estrus of rats on a high fat diet, *Proc. Natl. Acad. Sci. U.S.A.,* 72, 4172, 1975.
50. **Frisch, R. E. and McArthur, J.,** Menstrual cycles: fatness as a determinant of minimum weight for height necessary for their maintenance or onset, *Science,* 185, 949, 1974.
51. **Cole, P. and Cramer, D.,** Diet and cancer of endocrine target organs, *Cancer,* 40, 434, 1977.
52. **Petrakis, N. L., Gruenke, L. D., and Craig, J. C.,** Cholesterol and cholesterol epoxides in nipple aspirates of human breast fluid, *Cancer Res.,* 41, 2563, 1981.

53. **Gray, G. E., Pike, M. C., and Henderson, B. E.,** Breast cancer incidence and mortality rates in different countries in relations to known factors and dietary practices, *Br. J. Cancer,* 39, 1, 1979.
54. **Phillips, R. L.,** Role of life-style and dietary habits in risk of cancer among Seventh-Day Adventists, *Cancer Res.,* 35, 3513, 1975.
55. **Miller, A. B., Kelly, A., Choi, N. W., Matthews, V., Morgan, R. W., Munan, L., Burch, J. D., Feather, J., Howe, G. R., and Jain, M.,** A study of diet and breast cancer, *Am. J. Epidemiol.,* 107, 499, 1978.
56. **Howe, G. R.,** in preparation, 1985.
57. **Marshall, J. and Graham, S.,** Use of two observations to reduce distortion in risk assessment, *Am. J. Epidemiol.,* 114(Abstr.), 443, 1981.
58. **Graham, S., Marshall, J., Mettlin, C., Rzepka, T., Nemoto, T., and Byers, T.,** Diet in the epidemiology of breast cancer, *Am. J. Epidemiol.,* 116, 68, 1982.
59. **Lubin, J. H., Burns, P. E., Blot, W. J., Ziegler, R. G., Lees, A. W., and Fraumeni, J. F., Jr.,** Dietary factors and breast cancer risk, *Int. J. Cancer,* 28, 685, 1981.
60. **Nomura, A., Henderson, B. E., and Lee, J.,** Breast cancer and diet amoung the Japanese in Hawaii, *Am. J. Clin. Nutr.,* 31, 2020, 1978.
61. **Kolonel, L. N., Nomura, A. M. Y., Hinds, M. W., Hirohata, T., Hankin, J. H., and Lee, J.,** Role of diet in cancer incidence in Hawaii, *Cancer Res.,* 43, 2397s, 1983.
62. **Hiatt, R. A., Friedman, G. D., Bawol, R. D., and Ury, H. K.,** Breast cancer and serum cholesterol, *J. Nat. Cancer Inst.,* 68, 885, 1982.
63. **Rang, E. H., Kinlen, L. J., and Herman-Taylor, J.,** A study of cancer of the breast and other sites in woman after total colectomy and ileostomy for non-malignant disorders, *Lancet,* 1, 1014, 1983.
64. National Academy of Sciences, Diet, nutrition and cancer, Committee on Diet, Nutrition, and Cancer, Assembly of Life Sciences, National Research Council, National Academy Press, Washington, D.C., 1982.
65. **Miller, A. B.,** Approaches to the control of breast cancer, in *Understanding Cancer. Clinical and Laboratory Concepts,* Rich, M. A., Hager, J. C., and Furmanski, P., Eds., Marcel Dekker, New York, 1983, 3.
66. National Academy of Sciences, Diet, nutrition and cancer, Directions for research, Committee on Diet, Nutrition, and Cancer, Commission on Life Sciences, National Research Council, National Academy Press, Washington, D.C., 1983.
67. **Wynder, E. L., and Cohen, L.,** A rationale for dietary intervention in the treatment of postmenopausal breast cancer patients, *Nutr. Cancer,* 3, 195, 1982.
68. **Newman, S. C. and Miller, A. B.,** Diet and survival from breast cancer, in preparation, 1985.
69. **deWaard, F.,** Nutritional epidemiology of breast cancer: where are we now, and where are we going?, *Nutr. Cancer,* 4, 85, 1982.
70. **Kalache, A.,** personal communication, 1982.

Chapter 6

DIETARY FAT AND MAMMARY CANCER

Leonard A. Cohen

TABLE OF CONTENTS

I. INTRODUCTION

As a rule, discoveries relating environmental factors to human disease have had their origins in clinical or epidemiological observations, and only later were laboratory animal models developed to confirm these observations. For example, Lind's discovery of scurvy was made long before vitamin C was crystallized as was Snow's discovery of cholera before its bacterial etiology was established. However, in the case of dietary fat and mammary carcinogenesis, the initial observations occurred in a laboratory setting and over a decade passed before these findings were confirmed by epidemiological studies (see Chapter 5). In light of this unique sequence of events, it is appropriate to review not only the current state of experimental studies, but also the historical development of this fascinating and complex field.

Development of knowledge concerning the relationship between dietary fat and mammary carcinogenesis can be arbitrarily divided into three historical phases: (1) establishment of the fundamental tenet that dietary fat is an important determinant of mammary tumor development, (2) demonstration that fat exerts its effects primarily on the promotion phase of carcinogenesis, and (3) elucidation of mechanisms. In the following, each of these three stages will be traced, the current state of knowledge described, and future directions in research outlined.

A. Early Studies

In a sense, the past 25 years of research in this field can be viewed as a footnote to the seminal work of Albert Tannenbaum and colleague Herbert Silverstone. In their comprehensive 1953 review[1] (which itself had over 150 references) the basic principles of experimental nutritional carcinogenesis were laid down and described in detail. Based on a series of experiments using the spontaneous mouse mammary tumor model, it was established that (1) animals fed a high-fat (HF) diet, (12% (wt/wt) hydrogenated cottonseed oil) exhibited tumors earlier and with increased frequency than animals fed a low-fat (LF) diet (3% hydrogenated cotton seed oil) (Figure 1); (2) that the effect was not based on caloric intake; and (3) that the dose-response effect was nonlinear with a plateau occurring in the vicinity of 16% fat.

In addition to the obvious need to carefully control the numbers of animals used and the duration of the experiment, Tannenbaum stressed the importance of regulating the dose of carcinogen given to an animal so as not to block out the essentially modifying effects of exogenous factors such as dietary fat. (Ideally, the amount of carcinogen administered should lie on the most sensitive portion of the carcinogen/tumor dose-response curve).

A distinction between diet per se and "digestible caloric value" was also drawn as was the dichotomy between quantitative and qualitative differences in dietary fat content. Tannenbaum was well aware that the type of fat may play an important role, not only with respect to the degree of saturation, but also with regard to chain length. He also foresaw the use of pure synthetic triglycerides in place of natural randomly mixed triglycerides.

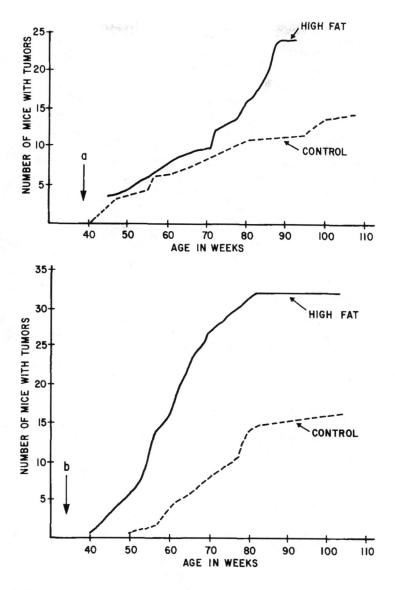

FIGURE 1.   Effects of a HF (12% hydrogenated cottonseed oil) diet on the formation of spontaneous breast tumors in female dba mice. Curves represent cumulative tumor incidence. (a) Diets instituted at 38 weeks. (b) Diets instituted at 24 weeks. Control diet consisted of 3% hydrogenated cottonseed oil. (From Tannenbaum, A., *Cancer Res.*, 2, 468, 1942. With permission.)

The importance of delineating the stage in the carcinogenic process where fat may exert its enhancing effect was emphasized as was the significance of body weight differentials in experimental groups as a potential confounding factor in dietary studies. Lastly, Tannenbaum suggested that the fat effect was due specifically to the triacylglycerol molecule and not to the glycerol moiety, phospholipids, or the nonsaponifiable fraction of dietary fat (sterols, including cholesterol). As this brief list indicates, Tannenbaum demonstrated not only an excellent grasp of the central experimental and conceptual problems inherent in nutrition and cancer studies but also a keen eye to the future development of the field.

B. The DMBA Model

The second major milestone in the development of our ideas concerning dietary fat and mammary carcinogensis was the pioneering work of Carroll using the DMBA tumor model.[2-6]

The discovery by Huggins[7] in the early 1960s that a single dose of the carcinogen 7,12-dimethylbenz(*a*)anthracene (DMBA) could induce mammary tumors in 100% of treated rats within 6 months opened the door to the quantitative study of exogenous factors which could suppress or enhance the carcinogenic process. The tumors produced were hormone responsive and resembled human breast cancers histologically and in other ways, with the important exception that they failed to metastasize. Most importantly the model provided an operational means to separate out those factors which act prior to, or during, initiation and those which act after the initial carcinogenic event (promotion). Using the DMBA model Carroll and colleagues published a series of papers in the late 1960s demonstrating that the primary effect of dietary fat was to enhance (or promote) the development of tumors in already initiated animals and that the amount of fat given prior to DMBA administration was of minimal importance.[3] Carroll also explored the effects of different types of fats and oils and was the first to show that polyunsaturated fats play a special role in the fat effect.[8]

Perhaps of equal importance, using the United Nations FAO Food Balance Tables[9] and Segi's Cancer Mortality Tables,[10] Carroll demonstrated a strong postive correlation between per capita "fat disappearance" and breast cancer mortality rates in a variety of countries around the world (see Chapter 5). In this way laboratory animal data and information from geographical epidemiology were linked for the first time and, as a result, a multidisciplinary approach to the etiology of breast cancer came into being.[11]

C. Elucidation of Mechanisms

The present era, that of exploration into mechanisms, was ushered in by Chan and Cohen in 1974 with the publication of a study involving the role of hormones and potential mediators of the fat effect.[12] In this study animals initiated with DMBA and fed high-fat (HF) and low-fat (LF) diets were treated with (1) 2-Br-α-ergocryptine, a dopamine agonist which suppresses pituitary prolactin secretion and (2) nafoxidine, an anti-estrogen which acts primarily to block the binding of estradiol to its specific target cell receptor. It was found that treatment with an antiprolactin drug eliminated the enhancing effect of a HF diet thereby suggesting that the HF effect may be mediated by prolactin via interaction with the hypo-thalamic-pituitary axis controlling prolactin secretion. The effects of the anti-estrogen, nafoxidine-HCl were less clear, since tumor incidence was lowered in both the HF and LF groups.

During the ensuing decade (1974 to 1984) a wide variety of papers have appeared dealing with possible mechanisms by which the HF effect is exerted in animal models. These will be discussed in detail in a later section.

II. CURRENT STATUS OF THE FIELD: A CRITICAL EVALUATION

A number of in-depth reviews of this field have already been published, the two most recent and comprehensive being a monograph published by the American Oil Chemists Society entitled *Dietary Fats and Health*[13] and the published proceedings of the workshop conference on Nutrition in Cancer Causation and Prevention.[14] The following review will therefore be illustrative rather than exhaustive, its primary purpose being to highlight current conceptual and experimental problems and point out gaps in our knowledge.

A. Dietary Fat as a Tumor Promoter: Operational vs. Biochemical Definitions of Promotion

While Carroll's original finding that the primary effect of dietary fat occurs at the post-initiation stage has been confirmed by other workers,[15-19] several recent studies suggest that high fat may also influence the initiation stage as well.[20,21] Dietary fat has been shown to alter the development of the mammary gland,[22,23] and to alter hepatic carcinogen metabolizing systems.[24-26] *In toto* however, the available evidence supports the idea that the primary effect of dietary fat is to enhance the development of already initiated mammary lesions.

It must be understood that the DMBA model provides what is essentially an operational definition of promotion. However, as recently discussed by Hick[27] a number of conceptual problems arise when the term "promotion" is applied to the fat effect. A "promoter" in the strictest sense is a substance which is not carcinogenic in itself but which is essential for the subsequent development of an initiated lesion into a tumor. With regard to dietary fat this is clearly not the case. Tumors will develop regardless of the amount of fat in the diet though fewer tumors develop in animals fed a LF diets. In addition, continuous exposure to a "classical" promoter is required for complete tumor development and the biological effects of a tumor promoter are, as a rule, reversible. A recent study[28] suggests that the fat effect may indeed be reversible but this remains to be demonstrated conclusively. Also a study by Dao and Chan indicated that tumor yield was directly proportional to the duration of HF feeding.[20] Overall, however the effects of dietary fat appear more like those reported for saccharin, cyclamates, and tryptophan metabolites in bladder cancer[29] and phenobarbitol and butylated hydroxytoluene (BHT) in hepatacellular carcinoma[30] than those of TPA (12-*O*-tetradecanoyl phorbol-13-acetate) in the mouse skin tumor system. For this reason the term "tumor enhancement" is often used synonomously with "promotion" when describing the fat effect.

A general biochemical definition of promotion remains to be established. The majority of studies on tumor promotion have been carried out with phorbol esters (TPA), agents which, though of great theoretical value, may have little practical relevance to tumor promotion in human cancer. Nonetheless, a good deal has been learned concerning the effects of TPA in cultured cells and in the two-stage mouse skin tumor model, particularly with regard to early membrane changes involving prostaglandin synthesis and polyamine metabolism, and the possible role of growth factors and their specific receptors.[31,32] Most recently TPA has been shown to activate membrane-bound protein kinase C (a phospholipid and Ca^{++}-dependent kinase) by acting as an analogue of the endogenous 1,2-diglyceride activator.[33,34] In summary while a multistage process of carcinogenesis no doubt occurs in the mammary gland, it is not necessarily identical to that in skin carcinogenesis, and the biochemical basis of mammary tumor enhancement by dietary fat remains to be elucidated.

B. Experimental Models, Endpoints, and Evaluation of Data Models

1. Tumor Models

The origins and characteristics of each tumor model must be taken into consideration when evaluating the results of dietary experiments. For example, the R3230AC transplantable rat mammary tumor arose "spontaneously" in a retired breeder. In contrast to the *N*-nitrosomethylurea (NMU) and DMBA-induced rat mammary tumors, the R3230AC tumor is inhibited rather than stimulated by exogenous prolactin administration.[35] In general, transplantable mouse mammary tumors are not hormone responsive[35] and spontaneous mouse mammary tumors contain associated virus particles[36] in contrast to rat mammary tumors, and human breast cancers. In addition, the histogenesis of the spontaneous mouse mammary tumor appears to differ from that of chemically induced rat tumors[37,38] and human breast cancers.[39] Moreover, because there is no definable "initiating" agent in the mouse model, operationally speaking one cannot distinguish between initiation and promotion. Similar considerations apply to the transplantable tumor model. The carcinogen-induced tumor has the disadvantage that a relatively high dose of carcinogen — several orders of magnitude higher than estimated human exposures — is administered to hasten tumor development and this adds an element of uncertainty with regard to the extrapolation of experimental results to humans. A major drawback to all the available models is their inability to metastasize to distant organ sites, a common characteristic of human breast cancers.[40]

2. Endpoints and Data Evaluation

Despite these problems, one of the most remakable aspects of the fat effect is its consistency from model to model and laboratory to laboratory. It has been demonstrated in spontaneous,[1,41] chemically induced,[2,15] X-ray-induced,[17] and transplantable mammary tumor models.[16] To the authors' knowledge only one transplantable tumor, the MT-W9B, does not respond to increased fat intake by the host.[42] Despite this consistency, it is crucial to recognize the differences inherent in each model and the endpoints measured since such differences are critical to the interpretation of results.

For example, when analyzing the effect of dietary fat on the development of chemically induced (autochthonous) mammary tumors four major endpoints can be measured: tumor incidence, tumor incidence, (no. animals tumors per total animals) tumor multiplicity (no. tumors per animal); tumor size and latency, or the time to first tumor. Rarely do HF diets significantly alter all four parameters and in some cases only one is changed in a significant fashion. For example, in Carroll's studies in the DMBA model the tumor-promoting effect of high-polyunsaturated fatty acid (PUFA) diets was evidenced primarily in terms of tumor number (or multiplicity); differences in latency or tumor incidence were negligible.[43] In a more recent study by Kalamegham and Carroll,[28] attempting to demonstrate the reversibility of the fat effect, a similar problem was encountered. Animals fed 20% sunflower seed oil throughout the experiment exhibited a final tumor count of 6.6 ± 3.9 tumors per tumor-bearing animals (mean $\pm$ SD); those fed 20% sunflower seed oil for 7 weeks and then transferred to a 10% butter diet exhibited a tumor count of 4.8 ± 3.3. No differences were reported in tumor incidence. While these data imply that the fat effect is reversible, one can readily see that they do not prove the point since the distributions of tumor count data from the two experimental groups show considerable overlap.

This is not a trivial point since as Lee[44] noted, tumor count data is subject to a number of uncontrollable variables. First, the number of tumors which develop per animal is a highly variable event even in animals given the same dose of carcinogen. Tumors often appear and disappear during an experiment and sometimes two or more adjacent tumors merge into one mass making counting difficult. Moreover, as seen in Figure 2, tumor count data do not fall into a normal or Gaussian distribution. Instead they fall into a logarithmic distribution, with most animals having none, some having one, or two tumors, followed by an asymptotic decrease in animals with higher tumor numbers. The higher the dose of initiating carcinogen the more the distribution is skewed towards the higher multiplicities; the lower the dose the more evenly the frequency distribution drops off. Hence, in order to test for significance between treatment groups using a common parametric test, such as analysis of variance tumor, count data should be subjected to a log (or square root) transformation which fulfills the assumption of a normal distribution and then tested for significance. Clearly, the above considerations point out the need for more careful attention to the evaluation of tumor multiplicity data. One reason for Carroll's failure to see significant differences in incidence and latency in the PUFA experiments may be that the dose of DMBA given was relatively high on the carcinogen dose-response curve (i.e., in a relatively insensitive zone) and as a result, the influence of the promoter was overwhelmed by the high initial carcinogenic stimulus.

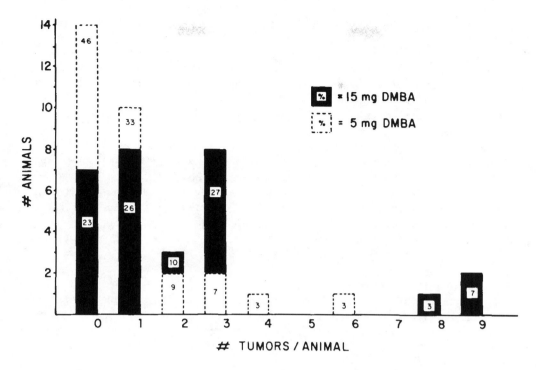

FIGURE 2.   Frequency distribution of tumor multiplicities in Sprague-Dawley rats initiated with high (15 mg) and low (5 mg) doses of DMBA. Tumors enumerated 20 weeks post-DMBA administration. (Adapted from Cohen, L. A., Polansky, M., Furuya, K., Reddy, M., Berke, B., and Weisburger, J. H., *J. Natl. Cancer Inst.*, 72, 165, 1984. With permission.)

An enhancing role for dietary PUFA has also been demonstrated in the transplantable mammary tumor model. A transplantable tumor differs markedly from an authochthonous one because as Foulds[45] pointed out some time ago, a transplantable tumor has not gone through the evolution in the host that an autochthonous tumor has. Moreover, it is during the progression of initiated lesion into a preneoplastic and *in situ* lesion into a tumor that tumor promoters appear to exert their effects and this progression is totally circumvented when a tumor fragment is implanted in a suitable host. Hence, studies using tumor implants are not, by present definitions, testing "promoting" effects of dietary fat; rather they are testing enhancement of growth of an already established tumor fragment.

Using the transplantable mammary tumor model, Abraham found that in certain tumors dietary linoleic acid (LA) C18:2 n-6 stimulated tumor growth while in others it did not.[23] Moreover, in one case diets with as little as 0.1% LA stimulated tumor growth to the same degree as diets containing 15% corn oil (7 to 8% LA), in marked contrast to results reported in chemically induced tumors.[43]

The transplantable tumor model also has the disadvantage that tumor weight is the endpoint and that the difference between "stimulated" and "basal" growth is often numerically small. For example, Abraham[46] found that mean ($\pm$ SD) tumor weight in 12 to 16 animals fed hydrogenated coconut oil was 0.69 $\pm$ 0.11 g whereas that in animals fed on a LA-containing diet was 1.58 $\pm$ 0.67 g. As the SDs indicate, there is a relatively small difference between the mean tumor weights of the two experimental groups, making fine distinctions between the two extremes difficult to evaluate.

Illustrative of a similar problem is a recent paper by Aylesworth et al.[47] using the chemically induced tumor model. In a study designed to test whether the duration of exposure to dietary fat determined the degree of tumor growth enhancement, the authors fed HF diets for varying periods of time and measured the standard endpoints. Interestingly, no differences were found in tumor incidence between experimental groups while significant differences were found primarily in terms of average tumor diameter and in some cases mean tumor number per animal and latency. Tumor diameter is a relatively crude measure subject to great random variability. As Kalamegham and Carroll[28] reported, analysis of frequency distributions showed no significant differences in the tumor size distribution patterns between different dietary groups. Moreover, the within-group and within-animal variability often are as great as, or greater than, the between-group variability.

Latent period — or the occurrence of tumors over time — is a useful measure since tumor promoters or inhibitors often act to increase or decrease the rate of tumor appearance. This measure has considerable relevance to the human disease, since delaying the appearance of a tumor by 5 years would have positive impact, particularly in the older postmenopausal age groups. Survival curve analysis provides a statistically sound means to assess differences in the rate of appearance of tumors in different experimental groups.[48,49] Some programs permit assessment of time to first tumor, time to second tumor, etc. allowing for even more detailed examination of the tumor data.

C. Body Weight Measurement

The importance of regular body weight measurements in a dietary experiment cannot be over emphasized. As seen in Figure 3 even after randomization into experimental groups there is considerable variability in animal weights. The range of weights in the group shown in Figure 3 varied from 128 to 173 g. This variability continues as the experiment progresses though there is a tendency to move away from the extreme ends of the distribution and towards the mean value. Between 4 and 8 weeks, the animal weight distributions show considerable overlap and this pattern continues as the animal growth rate plateaus over time.

While small differences in animal weights probably contribute little to the experimental outcome of dietary experiments, chronic underfeeding[50] and obesity[51,52] definitely influence final tumor incidence. However, a key attribute of the HF effect is that it occurs *independently* of obesity or underfeeding. This has been clearly demonstrated by Waxler et al.[51] and may help explain the ambiguous results of the many epidemiological studies attempting to relate obesity to increased risk of breast cancer (see Chapter 5).

D. Diets

When preparing HF and LF diets several basic rules must be complied with. The difference in fat content usually is varied from 5 to 20% (wt/wt) that is, 10 to 40% of total calories. The standard NIH-07 diet contains 4 to 5% fat. The 40% fat diet is chosen to mimic that of the HF "Western" diet,[53] and the 10% fat diet is designed to mimic the LF Japanese diet (as of the 1950s;[54] currently the Japanese diet is closer to 25% calories as fat.

In order to compensate for the increase in fat calories, carbohydrate is usually increased and protein held constant. This operation is based on Tannenbaum's report that changes in dietary carbohydrate content do not alter mammary tumor development.[1] However recent studies by Hoehn and Carroll[55] and Klurfeld et al.[56] indicate that the type of carbohydrate (simple vs. complex) does indeed influence tumor development. In particular it was shown that DMBA-treated rats fed carbohydrate in the form of sucrose or dextrose were more susceptible to mammary tumorigenesis than animals fed polysaccharides such as starch. While the reason for this is unclear at present, it is generally considered prudent to use a 1:4 ratio of monosaccharide to polysaccharide in studies where dietary fat is the key variable under examination.

Because of the increased caloric density of a HF diet — fat contains 9.2 kcals/g vs. 4.1 kcals/g for carbohydrate and protein — and because rodents, as a rule, eat isocalorically, animals on a HF diet consume less food than those on a LF diet and hence less vitamins

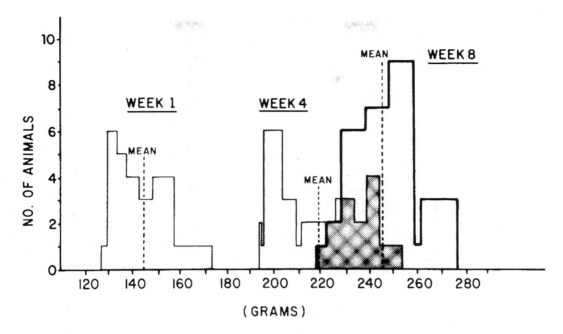

FIGURE 3.   Distribution of animal weights over time. Week 1 represents, 1 week post-DMBA treatment, i.e., day 62 of age. Darkened region represents area where weight distributions overlap. (Adapted from Cohen, L. A., Polansky, M., Furuya, K., Reddy, M., Berke, B., and Weisburger, J. H., *J. Natl. Cancer Inst.*, 72, 165, 1984. With permission.)

and minerals. As a consequence an "adjusted" HF diet was proposed[57] which in effect raised the levels of vitamins and trace elements in the HF diet to equalize comsumption of these nutrients in HF and LF diets. The adjusted HF diet assures that adequate amounts of vitamin E and A as well as the trace elements Sn and Zn, all of which are involved with controlling endogenous lipid peroxidation[58,59] and may influence mammary tumor development, are supplied (See Chapters 1 and 2 in Volume II).

In some cases, instead of compensating for increases in fat content by decreasing carbohydrate, an experimenter may hold carbohydrate constant and the quantity of nondigestible fiber decreased instead.[46] A potential problem with this approach lies in the fact that plant fibers can have profound effects on overall physiology by altering the absorption, metabolism, and excretion of nutrients. Depending on the type of fiber used, high fiber intake has also been associated with enhanced fecal excretion of Ca^{++}, Fe^{++}, Mg^{++}, and Zn^{++} due to the cation exchange properties of dietary fibers.[60,61]

As the above discussion indicates there are a number of methodological problems associated with the design of HF and LF diets. Essentially, it is impossible to change one dietary component without a compensatory change in another component. Hence an iron-clad monocausal proof is difficult to obtain. A LF diet, for example, could more accurately be called a LF-high complex-carbohydrate diet and a HF diet a HF-low complex-carbohydrate diet.

E. Elucidation of Mechanisms

Tannenbaum[1] suggested three basic mechanisms by which dietary fat may exert its effects. The first was that HF diets might modify the solubility, rate of metabolism, metabolite profile, or the total amount of a carcinogen reaching target cells in the mammary gland. The second envisioned that HF intake might modify the susceptibility of the target cells to the action of a carcinogen; and the third that it might modify the subsequent development of already initiated cells by altering either the immediate cell environment, the cell surface, or the surrounding stroma.

While the evidence is not conclusive, it appears unlikely that the first mechanism applies. Gammal et al.[62] for example, showed that the presence of DMBA and its metabolites in the mammary gland were not appreciably affected by HF diets. Later Chan et al.[15] using the direct-acting carcinogen, NMU, provided unambiguous proof that the HF effect could occur in the absence of any effects on host carcinogen metabolism. (DMBA, a large lipid soluble hydrocarbon requires prior activation by microsomal oxidases whereas NMU, a small water-soluble molecule, does not require prior host metabolism but yields an active methylating intermediated spontaneously under physiological conditions.) Moreover, since the fat effect can be demonstrated in spontaneous (1), X-ray-induced,[17] and transplantable tumors,[16] none of which involves carcinogen metabolism, it appears reasonable to assume that some other underlying mechanism is involved.

There is some evidence to support the second mechanism, an effect by dietary fat on mammary gland susceptibility to an "initiator". Huggins et al.[7] and later Russo et al.[38] have shown that there is a distinct stage in mammary gland development where it is most susceptible to the action of DMBA. This is around the time of puberty, and indirect evidence suggests that the same may hold true in humans.[62] However, despite some recent studies[20,21,52] suggesting that HF intake, prior to carcinogen intake, may alter mammary tumor development, there is no definitive evidence to suggest that HF diets significantly alter the development of the mammary gland, or its susceptibility to initiating agents *vis a vis* LF diets.

As stated earlier, the bulk of available evidence suggests that dietary fat exerts its tumor-enhancing effects on the postinitiation events in mammary carcinogenesis. Various mechanisms have been proposed to explain this phenomena, and these can be broadly classified into two categories: direct and indirect (Figure 4). Direct mechanisms involve quantitative/qualitative changes in the fatty acid composition of membrane phospholipids due to fat intake creating alterations in membrane transport, permeability phenomena, the activities of membrane-bound enzymes, and the status of hormone receptors. Direct changes also may involve alterations in cellular prostaglandin levels (prostaglandins are biologically active derivatives of essential fatty acids (EFA)) and/or the metabolism of EFA to prostaglandins. Indirect mechanisms involve neuroendocrine effects, i.e., elevations in the levels of circulating hormones and changes in brain amine metabolism, depression of immune rejection phenomena, and changes in gut bacteria and their subsequent metabolism of fecal bile acids to estrogen-like compounds.

1. Direct Effects
a. Membrane Fluidity

It is well established that the lipid profile of the cell membrane is in dynamic equilibrium with the environment[64] and that the fatty acyl composition of mammary tumors is responsive to dietary lipid intake within certain ranges.[8,46,65,66] Moreover, these changes can have profound physiological consequences: as Sabine[64] succinctly stated, "cellular membranes do as their lipids tell them."

One of the best studied phenomena is changes in membrane fluidity. Though often used loosely, the term "membrane fluidity" actually refers to a change in the physiochemical state of the membrane as measured by NMR (nuclear magnetic resonance) or microcalorimetry and not to any specific biological effect. In general, the PUFA/SF (saturated fatty acid) ratio determines membrane fluidity, but the major regulatory contribution relates to the degree of saturation of phospholipid acyl chains. While there is little doubt that HF

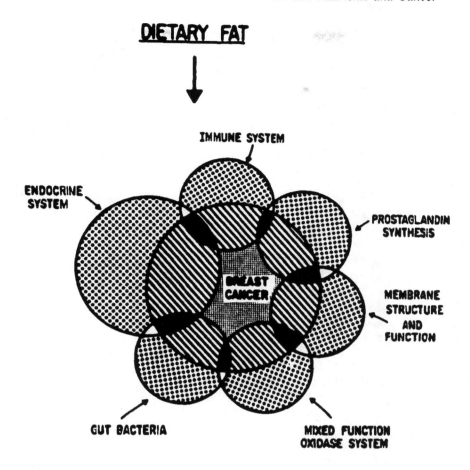

FIGURE 4. Biological mechanisms by which dietary fat may exert its stimulatory effects on breast cancer. Overlapping circles (*solid areas*) indicate that no single mechanism is mutually exclusive of any other mechanism. For example, an effect exerted via the immune system could also involve prostaglandin biosynthesis since immune reactions may be mediated by prostaglandins. Likewise, fat-induced changes in membrane structure could alter the activity of the membrane-bound mixed-function oxidase system. (From Cohen, L. A., *Cancer Res.*, 41, 3808, 1981. With permission.)

diets, especially PUFA-rich diets, may alter membrane structure and physiology it remains to be determined whether such changes are related to the HF effect on mammary tumorigenesis. For a comprehensive review of this area see Reference 67.

b. Prostaglandins

Effects of HF diets on prostaglandin (PG) metabolism and the possible association of PGs with mammary carcinogenesis are discussed in detail by Horrobin (see Chapter 7).

c. Endogenous Lipid Peroxides as Promoters

Oxidation of unsaturated fatty acyl groups via free radical reactions in biological membranes is a highly destructive process.[58] Cross-linking reactions between lipids and proteins, polymerization and polypeptide chain scission can cause loss of enzyme activity.[68] Membrane-bound cholesterol molecules can be metabolized to cholesterol epoxides which exhibit mutagenic activity in some systems,[69-71] although not in others.[7,73] In this context it is of interest that Petrakis et al.[74] reported that 17 of 37 healthy women exhibited detectable levels of cholesterol epoxide in their breast duct fluids. King et al.[19] using the antioxidant BHT reported a decrease in tumor incidence in animals fed HF diets containing the antioxidant BHT; paradoxically the effect was more pronounced in animals fed high-saturated fat than those fed high PUFA diets. However, in this experiment BHT was given before DMBA administration and since antioxidants can interfere with carcinogen metabolism it is uncertain whether BHT blocked the HF effect by acting to inhibit lipid peroxidation reactions or by accelerating carcinogen inactivation.[75] In conclusion, despite a large body of evidence that lipid peroxidation is related to a variety of pathological conditions, including carcinogenesis,[68] there is no compelling evidence at present to suggest that the fat effect is mediated by lipid peroxides or their breakdown products.

d. Hormone Receptors

Another potentially important membrane-related effect involves changes in the numbers and/or affinity of specific hormone receptors. In the mammary tumor model interest has focussed primarily on the prolactin receptor. This no doubt is because prolactin is a known promoter substance in a number of experimental mammary tumor systems[35] and because prolactin (PRL) receptors have been demonstrated in the NMU, DMBA tumors,[76-78] as well as in human breast cancers.[79] The rationale behind this approach is that the growth-promoting influence of a hormone is the product of at least two factors: (1) the amount of circulating hormone and (2) the number of target cell receptors available to bind and internalize the hormone. Theoretically, therefore elevation of the amount of PRL receptors expressed on the surface of a preneoplastic mammary cell could, in the presence of basal levels of circulating prolactin have the same net effect as elevated blood prolactin levels.

It is well known that the membrane-bound prolactin receptor is labile and can be induced both in mammary and hepatic tissue by administration of prolactin.[65,80,81] Prolactin receptors also appear to be regulated by dietary fat and particularly dietary LA. Cave et al.[65,76] reported elevated prolactin-binding capability in mammary tumors from animals fed HF diets. Interestingly this effect was associated with an increase in the LA content of tumor microsomes. Since LA is a major precursor of PGs,[82] these results suggest that the effect of HF diets on prolactin receptors may be mediated by tumor PGs.

e. Cell Culture Studies: Direct Cellular Effects of FFA (Free Fatty Acids)

The direct effects of dietary fat can also be tested by the use of in vitro systems. Cultured cells represent a valuable living model in that they provide a unique means to separate out host-mediated from direct effects of biologically significant agents. The disadvantages of cell culture systems are well known[83] and need not be discussed in detail here. They include the fact that cultured cells are usually grown in heterologous serum-containing medium or a nonphysiological mixture of growth factors, are subjected to selective processes in vitro, which do not exist in vivo, and must grow in the absence of normal tissue architecture and a circulation.

Most work in this area has involved the addition of specific fatty acids to the cell growth medium, either as a salt or bound to albumin, and measurement of specific indices of cell growth or differentiation. The studies by Kidwell et al.[84] and Wicha et al.,[85] for example, indicated that isolated primary cultures derived from DMBA tumors responded more to PUFA than to SF in terms of cell proliferation. Using a different culture system, Dulbecco et al.[86] reported that SF but not PUFA-induced "dome" formation (an index of differentiation in epithelial cells).

A problem with in vitro studies is that the precise lipid requirements of cultured mammary tumor cells are not known. It is known, however, that cells grown in vitro gain their lipid from the medium serum supplement in the form of free fatty acids (FFA) bound to serum albumin.[87] Very little is obtained via serum lipoproteins. Moreover, most studies have focussed on cells derived from already established tumors, whereas the promoting effect of

HF diets in vivo appears to be exerted primarily on the earlier preneoplastic stages of tumor development. The recent isolation of a "preneoplastic" mouse mammary tumor line by Medina[88] may provide a way out of this impasse. Probably the most valuable information to come out of these studies is that cultured mammary cells respond differently to different fatty acyl groups, particularly with regard to their degree of saturation and that the direction of change is consistent with that reported in in vivo feeding studies. The author and Karmali[89] have demonstrated that cultured rat mammary tumor cells can synthesize PGs from EFA precursors present in the growth medium and that when this synthesis is blocked cell growth is retarded; however more work is clearly in order, particularly with regard to the cellular and subcellular events involved in tumor cell EFA metabolism. Spector et al.[87] and Stubbs and Smith[67] have published detailed reviews of this research area.

The use of nonmammary cell culture systems to test for the effect of FFA on molecular "indices" of promotion is another potentially valuable approach to an understanding of the fat effect. Based on studies using TPA and other potential tumor promoters, such as saccharin, Trosko[90,91] has proposed that a breakdown in intercellular communications may be a key step in tumor promotion. Recently, Aylesworth et al.[47] reported that unsaturated fatty acids selectively blocked metabolic cooperation between Chinese hamster V79 cells while saturated fatty acids did not. However, oleic acid (C18:1 n-9) and palmitoleic acid (C16:1 n-9) also blocked metabolic cooperation suggesting that this effect is not mediated by PGs.

f. Effects of FFA on Intracellular Second Messenger Production

Other possible direct effects of dietary fatty acids involve interactions between specific fatty acids and guanylate cyclase (GC), which catalyzes to conversion of GTP to cyclic 3'-5' GMP, (cGMP).[92] Changes in the intracellular concentrations of second messengers such as cGMP and cyclic 3'-5'-AMP (cAMP) are of importance since the biological expression of a number of first messengers (hormones and neurotransmitters) is mediated by these second messengers.[92] Moreover, aberrant cAMP metabolism has been associated with expression of the malignant phenotype and differentiative functions in a number of tumor cell types.[93]

Studies in platelets using a soluble GC preparation indicate that the structural determinants required for fatty acids to serve as activators of the cyclase were almost identical to those required of fatty acids to serve as substrates for the fatty acid cyclooxygenase.[92] On the other hand, studies using crude fibroblast homogenates indicate that a membrane-associated GC can be stimulated by both saturated and unsaturated fatty acids.[94] Although the role of cyclic 3'-5' GMP in mammary tumor cell physiology is uncertain, these studies suggest that further studies on the influence of dietary fat on second messengers may be a fruitful area for further research.

2. Indirect Effects

a. Endocrine and Neuroendocrine Effects

i. Perturbations in Hormone Secretion Patterns

Perhaps the most intensively studied hypothetical mechanism involves perturbations in the endocrine system, specifically prolactin secretion induced by HF diets. This is primarily because most mammary tumors are hormone dependent at some point in their histogenesis,[95] including human breast cancers, and because prolactin, in particular, has long been known as a tumor promoter in mammary carcinogenesis.[35] The role of prolactin in human breast cancer, originally believed to be minimal, now appears to be better established.[96] Indirectly supporting a role for prolactin mediation is the fact that prolactin acts as a liporegulatory hormone in a number of vertebrate species.[97] As noted earlier, treatment of animals with the prolactin-suppressing drug 2-Br-α-ergocryptine effectively eliminated the HF effect.[12] Moreover, several early studies reported that immunoassayable serum prolactin concentrations were elevated in animals fed HF diets.[98,99] However, subsequent studies using more sophisticated methods involving cannulation (with multiple repeated sampling to follow the pulsitile release pattern of prolactin) have shown quite conclusively that HF diets do not change the amplitude or shape of the circadian rhythm of immunoreactive prolactin secretion by the pituitary (see References 100 and 101 for a complete review of this subject).

It should be noted that all of these studies have been based on radioimmunoassay (RIA) techniques and therefore measure only those circulating prolactin molecules which are immunoreactive. Recent studies, however, have shown that one can distinguish between immunoreactive and bioactive prolactin, and that prolactin can no longer be regarded as a single hormone but rather as a family of isohormones differing from one another in terms of size and charge and possibly bioactivity.[102-104]

These studies suggest that the molecular forms of prolactin vary and that this variation may have important functional significance. With regard to mammary cancer, Mittra has reported a "cleaved" 18 K form of prolactin in rats which has potent mitogenic activity on mammary epithelial cells whereas the major monomeric 22 K species does not.[105,106] Sinha and Gilligan[107] compared prolactin in female mice of a high (C3H/St) and a low (C57BL/St) mammary tumor strain and found that endogenous immunoreactive prolactin was found mainly in the "big" (polymeric) form in incubated pituitary glands of high incidence mice whereas it was present only in the "little" (monomeric) form in the low incidence strain. On the basis of this and other data the authors suggested that circulating monomeric prolactin in the high incidence strain existed largely in a nonimmunoreactive form.

The results of these studies suggest that prolactin can be secreted in forms not recognized by conventional RIA techniques and conversely that there are conditions under which prolactin can be recognized by RIA but cannot be detected by bioassay techniques. Accordingly, one can explain the 2-Br-α-ergocryptine data by postulating that dietary fat changes the type of prolactin secreted by the pituitary and that this change cannot be detected by conventional RIA techniques. This implies that a different prolactin variant (possibly Mittra's cleaved prolactin) may be secreted under HF conditions.

ii. HF Diets and Brain Amine Metabolism

Various hypothalamic amines are involved in the regulation of prolactin secretion by the pituitary gland.[108] Moreover, research over the past decade has shown that diet-induced plasma fluctuations in neurotransmitter precursors are important parts of feedback systems which control many brain functions.[109,110] The rate of synthesis of neurotransmitters, such as serotonin and acetylcholine, depend to a great extent on dietary precursor availability. Uptake of the essential amino acid tryptophan into neurons for example, initiates the synthesis of serotonin. Likewise, tyrosine uptake is closely linked to the synthesis of catecholamines such as dopamine which is a key regulator of prolactin secretion by the pituitary.[108] (Parenthetically, there is a striking similarity in the requirement for precursor essential amino acids for neurotransmitter synthesis and that demonstrated for precursor EFA requirements for PG biosynthesis. Both represent examples of the role of "essential" dietary nutrients in regulating the production of short-lived molecules which carry metabolic information from cell to cell).

Plasma-free tryptophan (and brain tryptophan) reportedly increases in animals fed HF diets.[110] This appears to be due to the displacement of albumin-bound tryptophan by the rise in serum EFA. While a number of methodological problems with these experiments remain to be clarified,[111-113] these findings suggest that HF diets may alter hypothalamic monomine synthesis by a mechanism involving displacement of albumin-bound amino acid precursors by circulating FFA. Such studies, based on differential competition for binding sites on serum albumin by essential amino and fatty acids, may provide a novel approach to our understanding of the fat effect.

b. Dietary Fat and Impaired Immunocompetence

A number of studies suggest that the tumor-promoting effect of high PUFA diets may be exerted indirectly by impairing the immune rejection response.[114,115] With regard to mammary cancer it has been demonstrated that animals fed HF diets have altered lymphocyte function. Kollmorgan et al.[116] reported that blastogenesis of spleen lympocytes in response to concanavalin A stimulation was suppressed in rats fed a high PUFA diet. However another study by the same group showed that the use of methanol extracts of *Bacillus Calmette-Guerin* were unable to reverse the tumor-enhancing effect of HF diets.[117] A possible connection between PUFA and immunosuppression via PG synthesis has been proposed[118] (see Chapter 7) but definitive proof of an immune mechanism for the HF effect has so far evaded the best efforts of investigators.

c. Dietary Fat, Gut Flora, Bile Acid Metabolism

An hypothesis involving the action of diet on gut flora has been proposed by Hill et al.,[119] according to which a HF diet enhances breast cancer by altering the composition and metabolic activity of the intestinal microflora and possibly also the bile acid substrates in the gut lumen. As a result of these changes, it is postulated, aromatization of the bile acid nucleus to an estrogen-like compound takes place and breast cancer cells are stimulated to proliferate. The evidence for this mechanism comes primarily from in vitro studies which indicated that the feces of individuals eating a standard Western diet tend to have a higher proportion of metabolically active anaerobic bacteria with the capacity to introduce double bonds into the A-ring of bile acids than do the feces of Africans.[119] Partially supporting the Hill hypothesis is a case-control study by Murray et al.[120] which showed that women with established breast cancer excrete significantly less fecal bile acids and conversely have significantly more bacterial nuclear dehydrogenation activity in their feces than controls.

The question of precisely which estrogens are synthesized in the gut and how much actually enters the circulation in vivo remain to be answered. Also, experiments in rodent models comparing the plasma estrogen profiles of germ-free and conventional animals on HF and LF diets would provide some of the information necessary to substantiate this hypothesis. A more fundamental problem with this hypothesis lies in the fact that there is little data to support the contention that elevated estrogen levels are related to either the HF effect[121] or increased risk for breast cancer.

In summary, while evidence on the whole remains fragmentary and circumstantial, the weight of available data favors a mechanism involving either neuroendocrine or prostaglandin mediation of the fat effect. Mediation by the immune system remains highly circumstantial as does a mechanism involving fecal bile acid metabolism.[122]

III. FUTURE STUDIES

Future trends point in four basic directions: investigations into the effects of special or altered fats, two-factor studies, further explorations into mechanisms, and analysis of the influence of dietary fat on the metastastic process.

A. Analysis of Special or Altered Fats

In addition to the commonly consumed plant seed oils such as corn, safflower and sunflower oils, which contain high levels of the principle EFA, linoleic acid (*cis, cis* C18:2 n-6) a number of different edible oils containing unique fatty acid profiles are currently under study. These include olive oil, *trans* oils, (principally elaidic acid), medium chain triglycerides (MCT), and marine oils.

1. Olive Oil

Olive oil, for example, consists of over 75% oleic acid (*cis* C18:1 n-9) and has been shown to lack tumor-promoting action in the NMU-induced mammary tumor model.[123] Also of interest, olive oil is the only edible oil which is cold-pressed rather than chemically processed prior to consumption.[124] In this regard, it is intriguing that in countries such as Italy, Greece, and Spain, where olive oil consumption is high, and as a consequence red cell and platelet levels of oleic are high (125), breast cancer mortality rates are intermediate with regard to those of the higher-risk Northern European Countries and low-risk Japan.[54] In an earlier study, Carroll[4] reported that olive oil exhibited tumor-promoting properties similar to that of corn oil, in contrast to our more recent studies. The reasons for this inconsistency are unclear but may involve differences in the country of origin of the oil and the tumor model used. Nonetheless, in light of the experimental and epidemiological findings and the question of processed vs. unprocessed oil, further investigation into the special effects of olive oil are merited.

2. Trans Oils

The potential health hazards of *trans* geometrical isomers of fatty acids has been under debate for a number of years.[126] It has been estimated that as much as 15% of fat intake in the U.S. consists of *trans* isomers particularly elaidic acid (*trans* C18:1 n-9). A report by Enig et al.[127] suggested that the upward trends in breast cancer incidence rates in the U.S. over the past 30 years closely follow the upward trend in *trans* acid consumption following the second World War, particularly in the form of margarines. While a recent report by Selenskas et al.[128] casts doubt on this relationship, these studies do suggest that geometrical, and possibly positional isomers of fatty acids merit further study with respect to their effects on mammary tumorigenesis.

3. Medium Chain Triglycerides (MCT)

MCT are a unique class of lipids derived from coconut oil. They consist mainly of C8:0 (65 to 75%) and C10:0 (25 to 35%), fatty acids. Although lipid in nature, MCT are absorbed and transported by the body in a manner more characteristic of carbohydrates than lipids. Instead of entering the lymphatic system in the form of chylomicrons like long-chain fatty acids, they are absorbed directly by the intestinal epithelium and delivered to the liver via the hepatic portal system. Once absorbed medium chain fatty acids (MCFA) are rapidly oxidized to CO_2 and H_2O and are rarely stored in fat depots or incorporated into cell membranes.[129] In a recent study,[130] the author and colleagues found that MCT have little or no tumor-promoting effect in the NMU-tumor model. These studies indicate the importance of chain length and degree of saturation in the fat effect and hold out the possibility that inclusion of special lipids such as MCT in the diet may have beneficial effects in terms of the reduction of breast cancer risk.

4. Marine Oils

Widespread interest has arisen over the past few years with regard to the potential beneficial health effects of long-chain polyunsaturated n-3 fatty acids such as eicosapentaenoic acid (EPA) (C20:5 n-3) and docosahexaenoic acid (DHA) (C22:6 n-3) present in fish oils. This interest emerged from the observation that Greenland Eskimos exhibit low rates of coronary thrombosis and cancer despite the fact that they eat HF diets.[131] EPA was found to interfere with the biosynthesis of the classical (n-6) PGs and to slow blood-clotting time. Based on the above considerations, and on the fact that PGs have been implicated in the growth and dissemination of mammary tumors[132] and that PGs are derived from EFA, the possibility has arisen that fish oils may have a beneficial (nonpromoting) effect in mammary carcinogenesis. Two studies have appeared in the literature, one suggesting that Menhadin oil

(15% EPA) has a promoting effect[43] similar to conventional fats and the other that it does not.[133] Clearly this is an area of great practical and theoretical interest and should be vigorously pursued.

5. Thermally Oxidized Fats

Much of the fat consumed in our diet has been exposed to heat during processing and in food preparation. Moreover, because of the presence of heat and oxygen, fats may undergo physiochemical changes such as oxidation and polymerization. When thermally oxidized fats are absorbed and metabolized they can produce toxic effects in animals.[134] The deleterious effects appear to reside primarily in what is called the distillable nonurea adductable fraction of the heated fat, which includes cyclic hydroxy esters of fatty acids, and dimeric fatty acids.[135]

With regard to mammary cancer, it has been proposed that oxidized lipids may be concentrated in breast duct fluids and initiate neoplastic transformation,[74] but as of the present time no studies have been reported concerning the tumor-enhancing effects of thermally oxidized dietary fats.

Arguing against a significant role for heat-altered lipids in disease processes is the fact that their ingestion is limited by their characteristic rancid taste. However, when one considers that an estimated 500 million lb of fats and oils are used each year in the manufacture of potato chips alone in the U.S.,[136] quantitative assessment of the influence of thermally oxidized fats in a mammary tumor model would appear appropriate at this time.

B. Two-Factor Studies

Two factor studies essentially involve analysis of the interactions between two dietary components. As of the present most of these studies have involved the interaction between dietary fat and protein. Visek and Clinton[137] demonstrated that protein-fat interactions are highly dependent on the relative proportions of the two components in the diet. When, for example, dietary protein was fed at 8% of calories, varying the amount of fat did not appreciably affect tumor incidence. With protein at 16 or 32% of calories, however, increasing dietary fat increased tumor incidence. One problem with such studies is the fact that epidemiological studies indicate that countries with widely varying risks for breast cancer such as Japan and the U.S. have similar protein intakes (approximately 11% of calories).[54]

Interactions between dietary fat and vitamins E and A (β-carotene), or the trace element Se, may also prove of interest since dietary fat influences the absorption of fat-soluble vitamins and because these vitamins and trace elements are involved in antioxidant functions.[58]

C. Elucidation of Mechanisms

Clearly the mechanism underlying any biological phenomena need not be fully understood to know how to control it. For example, control of prolactin synthesis and prolactin-associated disorders is fairly well understood despite the fact that the mechanisms of prolactin action is still only poorly understood.[138] Nonetheless, elucidation of the fat effect will enhance overall understanding of the carcinogenic process and in particular tumor promotion by natural as opposed to semisynthetic agents, such as TPA.

D. Metastasis

A dietary intervention feasibility trial has been proposed and is currently underway to test whether a LF diet can reduce the recurrance rate in women who have already had a mastectomy (Chapter 9). Unfortunately, demonstration of such an effect in a laboratory animal model is not yet possible. This is because available autochthonous rodent mammary tumor models do not metastasize in any regular fashion in marked contrast to human breast cancers.[40,139] The only available metastasizing systems — cultured tumor cells and transplantable tumors — are highly artifactual in that they do not recreate the process of dissemination of cells from a solid tumor as it occurs in the human disease.

As noted, the available metastasizing systems that have been established are either cultured cells or tumor implants. A number of rodent mammary tumor lines have been established[140-142] but they only rarely metastasize when inoculated back into syngeneic animals and then metastasis is usually confined to the lung, whereas the most common site of human metastasis is bone.[139] In addition metastasis is usually via the hematogenous route rather than the lymphatic route which predominates in human breast cancer metastasis. Recently, however, a line has been isolated which is suitable for the study of lymphogenous metastasis[143] and this may prove useful in analysis of dietary fats effects on the metastatic process.

Studies by Bennett et al.[144] and Honn and Meyer[145] suggest that prostanoids, and by implication their dietary precursors, may be involved in the metastatic process; however much further work will be necessary in this critical area before any definitive statements can be made (see Chapter 7).

IV. CONCLUSION

To quote from Tannenbaum's 1953 review:

"Early nutrition-cancer research had to be content with fact-finding and this has been accomplished to a satisfactory degree. More and more now attention is being turned to studying the mode of action of those diet alterations which affect tumorigenesis."[1]

Ironically, despite the passage of over 30 years we are presently in a position not far removed from that described above by Tannenbaum. The reason for the snail-like progress is two-fold. First, despite the profound implications of Tannenbaum's original work (and those preceding him), it had remarkably little impact on the biomedical community. This was due primarily to the fact that the reigning paradigm of the day was that nutritional deficiency was the primary cause of nutrition-related diseases and not overnutrition. Second, it was not until the link between geographic pathology and dietary fat was established that what was originally viewed as a laboratory curiosity was elevated to a serious scientific pursuit.

A time-trend curve for publications dealing with dietary fat and experimental mammary cancer over the past 3 decades is plotted in Figure 5. No doubt a similar trend would be obtained if epidemiological studies were plotted in a similar manner. Note that a general upward trend can be traced beginning in 1974. It is indeed unfortunate that Albert Tannenbaum passed away before he could see the fruits of his efforts.

ACKNOWLEDGMENTS

Supported by Research Grant CA29602 from the National Cancer Institute in Bethesda, MD. The author thanks Ms. Janet A. Fox for assistance in the preparation of this manuscript.

IN MEMORIUM

This review is dedicated to the memory of Dr. Albert Tannenbaum (1901 to 1980). During 2 decades (1935 to 1955), Dr. Tannenbaum conducted a series of experiments at Michael Reese Hospital in Chicago on nutrition in relation to cancer in mice. These studies provided the foundation for a wide variety of clinical, epidemiological, and experimental studies

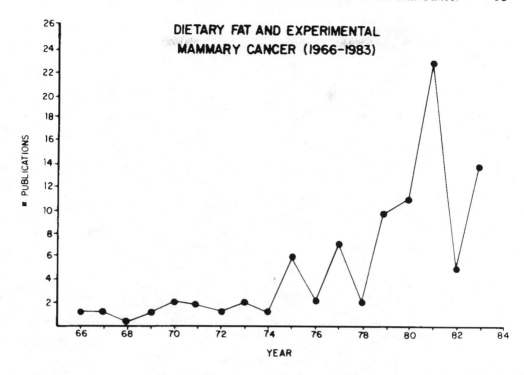

FIGURE 5.   Growth of literature on dietary fat and experimental mammary cancer (1966 to 1983). (Data obtained from Medlars (Medline) literature retrieval service of the National Library of Medicine.

linking nutrition and cancer, which culminated in the highly publicized report "Diet, Nutrition and Cancer", issued by the National Academy of Sciences in 1982. All of us owe Dr. Tannenbaum a debt of gratitude for his pioneering efforts in the field of nutritional carcinogenesis.

REFERENCES

1. **Tannenbaum, A. and Silverstone, H.,** Nutrition in relation to cancer, *Adv. Cancer Res.,* 1, 451, 1953.
2. **Carroll, K. K., Gammal, E. B., and Plunkett, E. R.,** Dietary fat and mammary cancer, *Can. Med. Assoc. J.,* 98, 590, 1968.
3. **Carroll, K. K. and Khor, H. T.,** Effects of dietary fat and dose level of 7,12-dimethylbenz(α)anthracene on mammary tumor incidence in rats, *Cancer Res.,* 30, 2260, 1970.
4. **Carroll, K. K. and Khor, H. T.,** Effects of level and type of dietary fat on incidence of mammary tumors induced in female Sprague-Dawley rats by 7,12-dimethylbenz(α)anthracene, *Lipids,* 6, 415, 1971.
5. **Carroll, K. K. and Khor, H. T.,** Dietary fat in relation to tumorigenesis, *Prog. Biochem. Pharmacol.,* 10, 308, 1975.
6. **Carroll, K. K.,** The role of dietary fat in carcinogenesis, in *Dietary Fats and Health,* Perkins, E. G. and Visek, W. J., Eds., American Oil Chemical Society, Champaign, Ill., 1983, 710.
7. **Huggins, C., Grand, L. C., and Brillantes, F. P.,** Mammary cancer induced by a single feeding of polynuclear hydrocarbons and its suppression, *Nature (London),* 189, 204, 1961.
8. **Hopkins, G. J., Kennedy, T. G., and Carroll, K. K.,** Polyunsaturated fatty acids as promoters of mammary carcinogenesis induced in Sprague-Dawley rats by 7,12-dimethylbenz(α)anthracene, *J. Natl. Cancer Inst.,* 66, 517, 1981.
9. Food Balance Sheets, 1975—1977 Average and per caput Food Supplies, 1961—1965 Average, 1967, to 1977, Food and Agricultural Organization, Rome, 1980.

10. **Segi, M., in collaboration with Hattori, H. and Segi, R.,** *Age-Adjusted Death Rates for Cancer for Selected Sites (A-Classification) in 46 Countries in 1975,* Segi Institute Cancer Epidemiology, Nagoya, Japan, 1980.

11. **Wynder, E. L. and Gori, G. B.,** Contribution of the environment to cancer incidence: an epidemiologic exercise, *J. Natl. Cancer Inst.,* 58, 825, 1977.

12. **Chan, P. and Cohen, L. A.,** Effect of dietary fat, antiestrogen and antiprolactin on the development of mammary tumors in rats, *J. Natl. Cancer Inst.,* 52, 25, 1974.

13. **Perkins, E. G. and Visek, W. J., Eds.,** *Dietary Fats and Health,* American Oil Chemists Society, Champaign, Ill., 1983.

14. Workshop conference on nutrition in cancer causation and prevention, *Cancer Res.,* 43(Suppl.), 2385, 1983.

15. **Chan, P. C., Head, J. F., Cohen, L. A., and Wynder, E. L.,** Influence of dietary fat on the induction of rat mammary tumors by *N*-nitrosomethyurea: associated hormone changes and differences between Sprague-Dawley and F344 rats, *J. Natl. Cancer Inst.,* 59, 1279, 1977.

16. **Hillyard, L. A. and Abraham, S.,** Effect of dietary polyunsaturated fatty acids on growth of mammary adenocarcinomas in mice and rats, *Cancer Res.,* 39, 4430, 1979.

17. **Silverman, J., Shellabarger, C. J., Holtzman, S., Stone, J. P., and Weisburger, J. H.,** Effect of dietary fat on x-ray induced mammary cancer in Sprague-Dawley rats, *J. Natl. Cancer Inst.,* 64, 631, 1980.

18. **Rogers, A. E. and Wetsel, W. C.,** Mammary carcinogenesis in rats fed different amounts and types of fat, *Cancer Res.,* 41, 3735, 1981.

19. **King, M. M., Bailey, D. M., Gibson, D D., Pitha, J. V., and McCoy, P. B.,** Incidence and growth of mammary tumors induced by 7,12-dimethylben(*a*)anthracene as related to the dietary content of fat and antioxidant, *J. Natl. Cancer Inst.,* 63, 657, 1979.

20. **Dao, T. L. and Chan, P-C,** Effect of duration of high fat intake on enhancement of mammary carcinogensis in rats, *J. Natl. Cancer. Inst.,* 71, 201, 1983.

21. **Ip, C.,** Ability of dietary fat to overcome the resistance of mature female rats to 7,12-dimethyl-benz(*a*)anthracene-induced mammary tumorigenesis, *Cancer Res.,* 40, 2785, 1980.

22. **Yang, L. S. and Rogers, A. E.,** Effect of dietary lard on mammary gland DNA synthesis and DMBA tumorigenesis in rats, *J. Nutr.,* 112, 31, 1982.

23. **Abraham, S. and Hillyard, L. A.,** Lipids, lipogenesis, and the effects of dietary fat on growth on mammary tumor model systems in mammary tumor model systems, in *Dietary Fats and Health,* Perkins, E. G. and Visek, W. J., Eds., American Oil Chemists Society, Champaign, Ill., 1983, 817.

24. **Nemoto, N. and Takayama, S.,** Effects of unsaturated fatty acids on metabolism of benzo(*a*)pyrene in an NADPH-fortified rat liver microsomal system, *Carcinogenesis,* 4, 1253, 1983.

25. **Hopkins, G. J. and West, C. E.,** Effect of dietary fats on pentobarbitone-induced sleeping times and hepatic microsomal cytochrome P-450 in rats, *Lipids,* 11, 736, 1976.

26. **Martin, C. W., Fjermestad, J., Smith-Barbaro, P., and Reddy, B. S.,** Dietary modification of mixed function oxidases, *Nutr. Rep. Int.,* 22, 395, 1980.

27. **Hicks, R. M.,** Pathological and Biochemical Aspects of tumour promotion, *Carcinogenesis,* 4, 1209, 1983.

28. **Kalamegham, R. and Carroll, K. K.,** Reversal of the promotional effect of high-fat diet on mammary tumorigenesis by subsequent lowering of dietary fat, *Nutr. Cancer,* 6, 22, 1984.

29. **Hicks, R. M.,** Effect of promotors on incidence of bladder cancer in experimental animal models, *Environ. Health Prospect.,* 50, 37, 1983.

30. **Peraino, C., Fry, R. J., Staffeldt, E., and Christopher, J. P.,** Enhancing effects of phenobarbitone and butylated hydroxytoluene on 2-acetyl-aminofluorene-induced hepatic tumorigenesis in the rat, *Food Cosmet. Toxicol.,* 15, 93, 1977.

31. **Slaga, T. J.,** Cellular and molecular mechanisms of tumour promotion, in *Cancer Surveys,* Vol. 2, Oxford University Press, 1983, 595.

32. **Marx, J. L.,** Do tumor promoters affect DNA after all?, *Science,* 219, 158, 1983.

33. **Sharkey, N. A., Leach, K. L., Blumberg, P. M.,** Competitive inhibition by diacylglycerol of specific phorbol ester binding, *Proc. Natl. Acad. Sci. U.S.A.,* 81, 607, 1984.

34. **Weinstein, I. B.,** News and views: tumor promoters, protein kinase, phospholipid and control of growth, *Nature (London),* 302, 750, 1983.

35. **Welsch, C. W. and Nagasawa, H.,** Prolactin and murine mammary tumorigenesis: a review, *Cancer Res.,* 37, 951, 1977.

36. **Sarkar, N. H., Fernandes, G., Telang, N., Kourides, I. A., and Good, R. A.,** Low calorie diet prevents the development of mammary tumors in C3H mice and reduces circulating prolactin level, murine mammary tumor virus expression, and proliferation of mammary alveolar cells, *Proc. Natl. Acad. Sci. U.S.A.,* 79, 7758, 1982.

37. **Haslam, S. Z. and Bern, H. A.,** Histopathogenesis of 7,12-dimethylbenz(*a*)anthracene-induced rat mammary tumors, *Proc. Natl. Acad. Sci. U.S.A.,* 74, 4920, 1977.

38. **Russo, J., Saby, J., Isenberg, W. M., and Russo, I. H.,** Pathogenesis of mammary carcinomas induced in rats by 7,12-methylbenz(*a*)anthracene, *J. Natl. Cancer Inst.,* 59, 435, 1977.
39. **Wellings, S. R., Jensen, H, M., and Marcum, R. G.,** An atlas of subgross pathology of the human breast with special reference to possible precancerous lesions, *J. Natl. Cancer Inst.,* 55, 231, 1975.
40. **Kim, U.,** Factors influencing metastasis of breast cancer, in *Breast Cancer Current Topics,* McGuire, W. L., Ed., Plenum Press, New York, 1979, 1.
41. **Tannenbaum, A.,** The genesis and growth of tumors. III. Effect of a high-fat diet, *Cancer Res.,* 2, 468, 1942.
42. **Ip, M. M. and Ip, C.,** Lack of effect of dietary fat on the growth and estrogen sensitivity of the MT-W9B transplantable mammary tumor, *Nutr. Cancer,* 3, 27, 1981.
43. **Carroll, K. K., Hopkins, G. J., Kennedy, T. G., and Davidson, M. B.,** Essential fatty acids in relation to mammary carcinogenesis, *Prog. Lipid Res.,* 20, 685, 1982.
44. **Lee, P. M.,** Statistical analysis of animal carcinogenesis experiments, in *Perspectives in Medical Statistics,* Bithall, J. F. and Coppi, R., Eds., Academic Press, New York, 1981, 86.
45. **Foulds, L.,** *Neoplastic Development,* Vol. 1, Academic Press, New York, 1969.
46. **Abraham, S. and Hillyard, L. A.,** Effect of dietary 18-carbon fatty acids on growth of transplantable mammary adenocarcinomas in mice, *J. Natl. Cancer Inst.,* 71, 601, 1983.
47. **Aylesworth, C. F., Jone, C., Trosko, J. E., Meites, J., and Welsch, C. W.,** Promotion of 7,12-dimethylbenz(*a*)anthracene-induced mammary tumorigenesis by high dietary fat in the rat: possible role of intercellular communication, *J. Natl. Cancer Inst.,* 72, 637, 1984.
48. **Gart, J. J., Chu, K. C., and Tarone, R. E.,** Statistical issues in interpretation of chronic bioassay tests for carcinogenicity, *J. Natl. Cancer Inst.,* 62, 957, 1979.
49. **Cohen, L. A., Polansky, M., Furuya, K., Reddy, M., Berke, B., and Weisburger, J. H.,** Inhibition of chemically induced mammary carcinogenesis in rats by short-term exposure to butylated hydroxytoluene (BHT): interrelationships among BHT concentration, carcinogen dose, and diet, *J. Natl. Cancer Inst.,* 72, 165, 1984.
50. **Tucker, M. J.,** The effect of long-term food restriction on tumours in rodents, *Int. J. Cancer,* 23, 803, 1979.
51. **Waxler, S. H., Brecher, G., and Beal, S. L.,** The effect of a fat enriched diet on the incidence of spontaneous mammary tumors in obese mice, *Proc. Soc. Exp. Biol. Med.,* 162, 365, 1979.
52. **Kritchevsky, D., Weber, M. M., and Klurfeld, D. M.,** Dietary fat *versus* caloric content in initiation and promotion of 7,12-dimethylbenz(*a*)anthracene-induced mammary tumorigenesis in rats, *Cancer Res.,* 44, 3174, 1984.
53. **Gortner, W. A.,** Nutrition in the United States 1900—1974, *Cancer Res.,* 35, 46, 1975.
54. **Hirayama, T.,** Epidemiology of breast cancer with special reference to diet, *Prev. Med.,* 7, 173, 1978.
55. **Hoehn, S. K. and Carroll, K. K.,** Effects of dietary carbohydrate on the incidence of mammary tumors induced in rats by 7,12-dimethylbenz(*a*)anthracene, *Nutr. Cancer,* 1, 27, 1979.
56. **Klurfeld, D. M., Weber, M. M., and Kritchevsky, D.,** Comparison of dietary carbohydrates for promotion of DMBA-induced mammary tumorigenesis in rats, *Carcinogenesis,* 5, 423, 1984.
57. **Newberne, P. M., Bieri, J. G., Briggs, G. M., and Nesheim, M. C.,** Control of diets in laboratory animal experimentation, *Anim. Res. News,* 21, A1, 1978.
58. **Logani, M. R. and Davies, R. E.,** Lipid oxidation: biologic effects and antioxidants — a review, *Lipids,* 15, 485, 1980.
59. **Bartoli, G. M., Bartoli, S., Galeotti, T., and Bertoli, E.,** Superoxide dismutase content and microsomal lipid composition of tumours with different growth rates, *Biochim. Biophys. Acta,* 620, 205, 1980.
60. **Anderson, J. W. and Chen, W-J-L,** Plant fiber carbohydrate and lipid metabolism, *Am. J. Clin. Nutr.,* 32, 346, 1979.
61. **Kritchevsky, D.,** Fiber, steroids, and cancer, *Cancer Res.,* 43(Suppl.), 24915, 1983.
62. **Gammal, E. B., Carroll, K. K., and Plunkett, E. R.,** Effects of dietary fat on the uptake and clearance of 7,12-dimethylbenz(*a*)anthracene by rat mammary tissue, *Cancer Res.,* 28, 384, 1968.
63. **McGregor, D. H., Land, C. E., Choi, K., Tokuoka, S., Liu, P. I., Wakabayashi, T., and Beebe, G. W.,** Breast cancer incidence among atomic bomb survivors Hiroshima and Nagasaki, 1950—69, *J. Natl. Cancer Inst.,* 59, 799, 1977.
64. **Sabine, J. R.,** Lipids in cancer — a functional or a fictional association?, *Trends Biochem. Sci.,* 8, 234, 1983.
65. **Cave, W. T. and Jurkowski, J. J.,** Dietary lipid effects on the growth, membrane composition, and prolactin-binding capacity of rat mammary tumors, *J. Natl. Cancer Inst.,* 73, 185, 1984.
66. **Chan, P. C., Ferguson, K. A., and Dao, T. L.,** Effects of different dietary fats on mammary carcinogenesis, *Cancer Res.,* 43, 1029, 1983.
67. **Stubbs, C. D. and Smith, A. D.,** The modification of mammalian membrane polyunsaturated fatty acid composition in relation to membrane fluidity and function, *Biochim. Biophys. Acta,* 779, 89, 1984.

68. **Ames, B. N.,** Dietary carcinogens and anti-carcinogens, *Science*, 221, 1256, 1983.
69. **Parsons, P. G. and Goss, P.,** Chromosome damage and DNA repair induced in human fibroblasts by UV and cholesterol oxide, *Aust. J. Exp. Biol. Med. Sci.*, 56, 287, 1978.
70. **Heiger, I.,** Carcinogenic activity of lipoid substances, *Br. J. Cancer*, 3, 123, 1949.
71. **Kelsey, M. I. and Pienta, R. J.,** Transformation of hamster embryo cells by cholesterol-α-epoxide and lithocholic acid, *Cancer Lett.*, 6, 143, 1979.
72. **Keena, T. W. and Patton, S.,** Cholesterol esters of milk and mammary tissue, *Lipids*, 5, 42, 1970.
73. **Smith, L. L., Smart, V. A., and Ansar, G. A. S.,** Mutagenic cholesterol preparations, *Mutat. Res.*, 68, 23, 1979.
74. **Petrakis, N. L., Gruenke, L. D., and Craig, J. C.,** Cholesterol and cholesterol epoxides in nipple aspirates of human breast fluid, *Cancer Res.*, 41, 2563, 1981.
75. **Wattenberg, L. W.,** Inhibition of chemical carcinogenesis, *J. Natl. Cancer Inst.*, 60, 11, 1978.
76. **Cave, W. T., Jr. and Erickson-Lucas, M. J.,** Effects of dietary lipids on lactogenic hormone receptor binding in rat mammary tumors, *J. Natl. Cancer Inst.*, 68, 319, 1982.
77. **Costlow, M. E., Buschow, R. A., and McGuire, W. L.,** Prolactin receptors and androgen-induced regression of 7,12-dimethylbenz(a)anthracene-induced mammary carcinoma, *Cancer Res.*, 36, 3324, 1976.
78. **Kledzik, G. S., Bradley, C. J., Marshall, S., Campbell, G. A., and Meites, J.,** Effects of high doses of estrogen on prolactin-binding activity and growth of carcinogen-induced mammary cancers in rats, *Cancer Res.*, 36, 3265, 1976.
79. **Beyrat, J. P., Dewailly, D., Djiane, J., Kelley, P. A., Vandewoole, B., Bonneterre, J., and Lefebre, J.,** Total prolactin binding sites in human breast biopsies, *Breast Cancer Res. Treat.*, 1, 369, 1981.
80. **Costlow, M. E., Buschow, R. A., and McGuire, W. L.,** Prolactin stimulation of prolaction receptors in rat liver, *Life Sci.*, 17, 1457, 1975.
81. **Posner, B. I., Kelley, P. A., and Friesan, H. G.,** Induction of a lactogenic receptor in rat liver: influence of estrogen and the pituitary, *Proc. Natl. Acad. Sci. U.S.A.*, 71, 2407, 1974.
82. **Van Dorp, D. A., Beethuis, R. K., Nugteren, D. H., and Vonkeman, H.,** Enzymatic conversion of all *cis* polyunsaturated fatty acid into prostaglandins, *Nature (London)*, 203, 839, 1964.
83. **Rothblat, G. H. and Cristofalo, V. J.,** Eds., *Growth, Nutrition, and Metabolism of Cells in Culture*, Vol. 3, Academic Press, New York, 1977.
84. **Kidwell, W. R., Monaco, M. E., Wicha, M. S., and Smith, G. S.,** Unsaturated fatty acid requirements for growth and survival of a rat mammary tumor cell line, *Cancer Res.*, 38, 4091, 1978.
85. **Wicha, M. S., Liotta, L. A., and Kidwell, W. R.,** Effects of free fatty acids on the growth of normal and neoplastic rat mammary epithelial cells, *Cancer Res.*, 39, 426, 1979.
86. **Dulbecco, R., Bologna, M., and Unger, M.,** Control of differentiation of a mammary cell line by lipids, *Proc. Natl. Acad. Sci. U.S.A.*, 77, 1551, 1980.
87. **Spector, A. A., Mathur, S. N., Kaduce, T. L., and Hyman, B. T.,** Lipid nutrition and metabolism of cultured mammalian cells, *Prog. Lipid Res.*, 19, 155, 1981.
88. **Medina, D. and Asch, B. B.,** Cell markers for mouse neoplasias, in *Cell Biology of Breast Cancer*, McGrath, C. M., Brennan, M. J., and Rich, M. A., Eds., Academic Press, New York, 1980, 323.
89. **Cohen, L. A. and Karmali, R. H.,** Endogenous prostaglandin production by established cultures of neoplastic rat mammary epithelial cells, *In Vitro*, 20, 119, 1984.
90. **Trosko, J. E., Dawson, B., Yotti, L. P., and Chang, C. C.,** Saccharin may act as a tumor promoter by inhibiting metabolic cooperation between cells, *Nature (London)*, 284, 109, 1980.
91. **Trosko, J. E. and Chang, C. C.,** An integrative hypothesis linking cancer, diabetes and atherosclerosis: the role of mutation and epigenetic changes, *Med. Hypoth.*, 6, 455, 1980.
92. **Goldberg, N. D. and Haddox, M. K.,** Cyclic GMP metabolism and involvement in biological regulation, *Ann. Rev. Biochem.*, 46, 823, 1977.
93. **Cho-Chung, Y. S.,** Mode of cyclic AMP action in growth control, in *Hormone Regulation of Mammary Tumors*, Vol. 2, Leung, B. L., Ed., Eden Press, Montreal, Canada, 1982, 155.
94. **Ichihara, K., El-Zayat, M., Mittal, C. K., and Murad, F.,** Fatty acid activation of guanylate cyclase from fibroblasts and liver, *Arch. Biochem. Biophys.*, 197, 44, 1979.
95. **Smithline, F., Sherman, L., and Kolodny, H. D.,** Prolactin and breast carcinoma, *New Engl. J. Med.*, 292, 784, 1975.
96. **Malarkey, W. B., Kennedy, M., Allred, L. E., and Milo, G.,** Physiological concentrations of prolactin can promote the growth of human breast tumor cells in culture, *J. Clin. Endocrinol. Metab.*, 56, 673, 1983.
97. **Meier, A. H.,** Prolactin, the liporegulatory hormone, in *Comparative Endocrinology of Prolactin*, Dellman, H. D., Johnson, T. A., and Klachko, D. M., Eds., Plenum Press, New York, 1977, 153.
98. **Chan, P. C., Didato, F., and Cohen, L. A.,** High dietary fat elevation of rat serum prolactin and mammary cancer, *Proc. Soc. Exp. Biol. Med.*, 149, 133, 1975.
99. **Ip, C., Yip, P., and Bernardis, L. L.,** Role of prolactin in the promotion of dimethylbenzanthracene-induced mammary tumors by dietary fat, *Cancer Res.*, 40, 374, 1980.

100. **Welsch, C. W. and Aylsworth, C. F.,** Enhancement of murine mammary tumorigenesis by feeding high levels of dietary fat: a hormonal mechanism?, *J. Natl. Cancer Inst.,* 70, 215, 1983.

101. **Dao, T. L. and Chan, P. C.,** Hormones and dietary fat as promoters in mammary carinogenesis, *Environ. Health Perspect.,* 50, 219, 1983.

102. **Klindt, J., Robertson, M. C., and Friesan, H. G.,** Episodic secretary patterns of rat prolactin determined by bioassay and radioimmunoassay, *Endocrinology,* 111, 350, 1982.

103. **Asawaroengchi, H. and Nicoll, C. S.,** Relationship among bioassay, radioimmunoassay and disc electrophoretic assay methods of measuring rat prolactin in pituitary tissue and incubation medium., *J. Endocrinol.,* 73, 301, 1977.

104. **Whitaker, M. D., Klee, G. G., Kao, P. C., Randall, R. V., and Heser, D. H.,** Demonstration of biological activity of prolactin molecular weight variants in human sera, *J. Clin. Endocrinol. Metab.,* 58, 826, 1984.

105. **Mittra, I.,** A novel "cleaved prolactin" in the rat pituitary. I. Biosynthesis, characterization and regulatory control, *Biochem. Biophys. Res. Commun.,* 95, 1750, 1980.

106. **Mittra, I.,** A novel "cleaved prolactin" in the rat pituitary. II. *In vivo* mammary mitogenic activity of its N-terminal 16K moiety, *Biochem. Biophys. Res. Commun.,* 95, 1760, 1980.

107. **Sinha, Y. N. and Gilligan, T. A.,** A "cleaved" form of prolactin in the mouse pituitary gland: identification and comparison of *in vitro* synthesis and release in strains with high and low incidences of mammary tumors, *Endocrinology,* 114, 2046, 1984.

108. **Quay, W. B.,** Biogenic amines in neuroendocrine systems: multiple sources, messages, targets and controls, *Tex. Rep. Biol. Med.,* 38, 87, 1979.

109. **Wurtman, R. J. and Fernstrom, J. D.,** Control of brain neurotransmitter synthesis by precursor availability and nutritional state, *Biochem. Pharmacol.,* 25, 1691, 1976.

110. **Anderson, G. H. and Johnston, J. L.,** Nutrient control of brain neurotransmitter synthesis and function, *Can. J. Physiol. Pharmacol.,* 61, 271, 1983.

111. **Badawy, A. A-B, Morgan, C. J., Davis, N. R., and Dacey, A.,** High-fat diets increase tryptophan availability to the brain: importance of choice of control diet, *Biochem. J.,* 217, 863, 1984.

112. **Cousins, C., Marsden, C. A., and Brindley, D. N.,** Feeding rats on diets rich in fat need not alter the concentrations of 5-hydroxytryptamine and 5-hydroxyindoleacetic acid in the brain, *Biochem. J.,* 206, 431, 1982.

113. **Brindley, D. N., MacDonald, I. A., and Marsden, C. A.,** High-fat diets need not increase tryptophan availability to the brain: importance of the choice of the control diet, *Biochem. J.,* 217, 865, 1984.

114. **Meade, C. J. and Mertin, J.,** Fatty acids and immunity, in *Advances in Lipid Research, Vol. 16,* Paoletti, R. and Kritchevsky, D., Eds., 1978, 127.

115. **Yoo, T-J, Kuo, C-Y, Spector, A. A., Denning, G. M., Floyd, R., Whiteaker, S., Kin, H., Kim, J., Abbas, M., and Budd, T. W.,** Effect of fatty acid modification of cultured hepatoma cells on susceptibility to natural killer cells, *Cancer Res.,* 42, 3596, 1982.

116. **Kollmorgan, G. M., Sansing, W. A., Lehman, A. A., Fischer, G., Loughley, R. E., Alexander, S. S., King, M. M., and McCay, P.,** Inhibition of lymphocyte function in rats fed high-fat diets, *Cancer Res.,* 39, 3458, 1979.

117. **Kollmorgan, G. M., King, M. M., Lehman, A. A., Fischer, G., Longley, R. E., Daggs, B. J., and Sansing, W. A.,** The methanol extraction residue of *Bacillus Calmette - Guerin* protects against 7,12-dimethylbenz(*a*)anthracene-induced rat mammary carcinoma *Proc. Soc. Soc. Exp. Biol. Med.,* 162, 410, 1979.

118. **Ceuppens, J. and Goodwin, J.,** Prostaglandins and the immune response to cancer, (review), *Anticancer Res.,* 1, 71, 1981.

119. **Hill, M. J., Goddard, P., and Williams, R, E. O.,** Gut bacteria and aetiology of cancer of the breast, *Lancet,* 2, 472, 1971.

120. **Murray, W. R., Blackwood, A., Calman, K. C., and Mackay, C.,** Faecal bile acids and clostridia in patients with breast cancer, *Br. J. Cancer,* 42, 856, 1980.

121. **Zumoff, B.,** The role of endogenous estrogen excess in human breast cancer (review), *Anticancer Res.,* 1, 39, 1981.

122. **Cohen, L. A.,** Mechanisms by which dietary fat may stimulate mammary carcinogenesis in experimental animals, *Cancer Res.,* 41, 3808, 1981.

123. **Cohen, L. A. and Thompson, D. O.,** Dietary fat, serum lipids, and the development of *N*-nitrosomethylurea-induced mammary tumors, *Fed. Proc.,* 43, 614, 1984.

124. Proc. 3rd Int. Congr. Biol. Value of Olive Oil, Int. Olive Oil Council, Madrid, Spain, 1980.

125. **Dougherty, R. M. and Iacono, J. M.,** Interrelationship of dietary fat and lipid composition of red blood cells and platelets, *Fed. Proc.,* 43, 865, 1984.

126. **Gottenbos, J. J.,** Biological effects of *trans* fatty acids, in *Dietary Fats and Health,* Perkins, E. G. and Visek, W. J., Eds., American Oil Chemical Society, Champaign, Ill., 1983, 375.

127. **Enig, M. G., Mann, R. J., and Keeney, M.,** Dietary fat and cancer trends — a critique, *Fed. Proc.,* 37, 2215, 1978.

128. **Selenskas, S. L., Ip, M. M., and Ip, C.,** Similarity between *trans* fat and saturated fat in the modification of rat mammary carcinogenesis, *Cancer Res.,* 44, 1321, 1984.

129. **Bach, A. C. and Babayan, V. K.,** Medium chain triglycerides: an update, *Am. J. Clin. Nutr.,* 36, 950, 1982.

130. **Cohen, L. A., Thompson, D. O., Maeura, Y., and Weisburger, J. H.,** The influence of dietary medium chain triglycerides on the development of N-methylnitrosourea-induced mammary tumors, *Cancer Res.,* 44, 5023. 1984.

131. **Dyerberg, J.,** Observations on populations in Greenland and Denmark, in *Nutritional Evaluation of Long-Chain Fatty Acids in Fish Oil,* Barlow, S. M. and Stansby, M. E., Eds., Academic Press, New York, 1982, 245.

132. **Karmali, R.,** Prostaglandins and cancer CA, *Cancer J. Clin.,* 33, 29, 1984.

133. **Jurkowski, J. J. and Cave, W. T., Jr.,** Dietary effects on a N-3 polyunsaturated lipid (Menhaden oil) on the growth and membrane composition of rat mammary tumors, *Proc. Am. Assoc. Cancer Res.,* 25, 211, 1984.

134. **Alexander, J. C.,** Biological effects due to changes in fats during heating, *J. Am. Oil Chem. Soc.,* 55, 711, 1978.

135. **Shue, G. M., Douglass, C. D., Firestone, D., Friedman, L., Friedman, L., and Sage, J. S.,** Acute physiological effects of feeding rats non-urea-adducting fatty acids (urea-filtrate), *J. Nutrition,* 94, 171, 1968.

136. **Chang, S. S., Peterson, R. J., and Ho, C-T,** Chemical reactions involved in the deep-fat drying of foods, *J. Am. Oil Chem. Soc.,* 55, 718, 1978.

137. **Visek, W. J. and Clinton, S. K.,** Dietary fat and breast cancer, in *Dietary Fats and Health,* Perkins, E. G. and Visek, W. J., Eds., American Oil Chemists Society, Champaign, Ill., 1983, 721.

138. **Rillema, J. A.,** Mechanism of prolactin action, *Fed. Proc.,* 39, 2593, 1980.

139. **Hortobagyi, G. N., Libshitz, H. I., and Seabold, J. E.,** Osseous metastases of breast cancer, *Cancer,* 53, 577, 1984.

140. **Kano-Sueka, T. and Hsieh, P.,** A rat mammary carcinoma *in vivo* and *in vitro:* establishment of clonal lines of the tumor, *Proc. Natl. Acad. Sci. U.S.A.,* 70, 1922, 1973.

141. **Cohen, L. A.,** Isolation and characterization of a serially cultivated neoplastic epithelial cell line from the N-nitrosomethylurea induced rat mammary adenocarcinoma, *In Vitro,* 18, 565, 1982.

142. **Bennett, D. C., Peachey, L. A., Durbin, H., and Rudland, P. S.,** A possible mammary stem cell line, *Cell,* 15, 283, 1978.

143. **Neri, A., Welch, D., Kawaguchi, T., and Nicholson, G. L.,** Development and biologic properties of malignant cell sublines and clones of a spontaneously metastasizing rat mammary adenocarcinoma, *J. Natl. Cancer Inst.,* 68, 507, 1982.

144. **Bennett, A., Berstock, D. A., Carroll, M. A., Stamford, I. F., and Wilson, A. J.,** Breast cancer its recurrence and patient survival in relation to tumor prostaglandins, *Adv. Prostaglandin Thrombox, Leukotr. Res.,* 12, 299, 1983.

145. **Honn, K. V. and Meyer, J.,** Thromboxanes and prostacyclin: positive and negative modulators of tumor growth, *Biochem. Biophys. Res. Commun.,* 102, 1122, 1981.

Chapter 7

THE ROLE OF ESSENTIAL FATTY ACIDS AND PROSTAGLANDINS IN BREAST CANCER

David F. Horrobin

TABLE OF CONTENTS

I. INTRODUCTION

The stimuli to investigate the relationship between dietary fat and breast cancer came from two directions. Animals fed high-fat diets showed increased development of mammary gland tumors.[1] In humans there was found to be a close correlation between mortality from breast cancer and total fat consumption on a country by country basis.[2] Over the past decade there has been an explosion of research in this area and the purpose of this review is to consider the evidence that essential fatty acids (EFAs) and their metabolites such as hydroxy-acids, peroxides, prostaglandins (PGs), and leukotrienes (LTs) may be involved in the fat effect.

For understandable reasons, work on animals has far exceeded in volume work on humans. Nevertheless, the limitations of animal work must constantly be borne in mind. We do not understand human breast cancer, a disease with a highly variable course occurring in genetically diverse populations. The idea that any of the animal mammary diseases which we study is a true "model" of the human disease is nothing more than a highly speculative assumption. We shall know whether or not any animal cancer is a true model only when both the human and the animal disease are fully understood. There is a considerable risk that animal models, using highly inbred strains with either a very high susceptibility to spontaneous tumors, or transplanted with aggresive tumors, or exposed to powerful carcinogens, may bear only a tenous relation to the human problem. (For further discussion see Chapter 6). Therefore the human and animal studies must be considered separately. Attempts will be made to correlate the two, but as will be seen later, the conclusions to be drawn sometimes seem diametrically opposed.

II. ESSENTIAL FATTY ACIDS

Essential fatty acids (EFAs) are polyunsaturated fatty acids (PUFAs) which are capable of correcting the deficiency disease which develops when animals are supplied a diet which except for its PUFA content contains all known essential dietary factors. All EFAs are PUFAs but many PUFAs are not EFAs. There are two main dietary EFAs which can be manufactured by plants but not animals. These are *cis:cis*-linoleic acid (18:2 n-6) and *cis:cis:cis*-α-linolenic acid (18:3 n-3). These are metabolized within the body by the pathways shown in Figure 1. There is universal agreement about the essentiality of the n-6 EFAs and their efficiency as EFAs tends to increase on moving from 18:2 to 20:4. There is much controversy about the essentiality of the n-3 EFAs since it is difficult to demonstrate any clinical abnormalities in either animals or humans deprived of n-3 EFAs. However the long chain n-3 EFAs are found throughout the body and in very large amounts in brain and retina. n-3 and n-6 EFAs appear to compete for the same enzyme systems with the n-3 compounds being preferentially metabolized.[3]

In order for a fatty acid to have EFA activity, all its double bonds must be in the *cis* configuration. Even one bond in the *trans* form leads to loss of EFA activity and to competitive inhibition of the metabolism of all *cis* EFAs. EFAs from the usual plant sources are all *cis*. Small amounts of *trans* fatty acids are formed in the rumen and so dairy products may contain *trans* fatty acids, usually at a level of well under 10% of the total fatty acids present. The majority of the *trans* acids in the modern diet are formed during vegetable oil processing. *Trans* bonds are usually more stable and so are deliberately indroduced with the aim of ensuring longer shelf life. *Trans* acid content of foods has increased dramatically since the 1920s and they are found in a wide variety of processed foods, candies, and bakery products as well as in foods which are more obviously related to vegetable oils.[4,5]

EFAs, like other fats, can be used to supply energy but that is not the basis of their essentiality. EFAs are vital components of all membranes, determining membrane fluidity

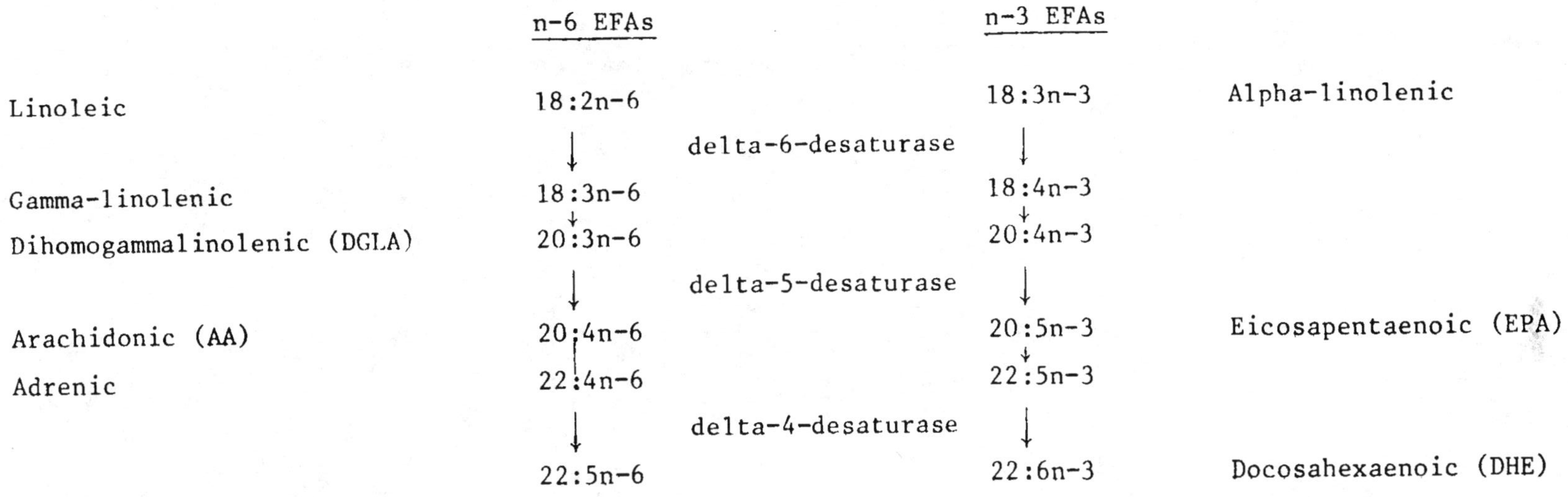

	n-6 EFAs		n-3 EFAs	
Linoleic	18:2n-6		Alpha-linolenic	18:3n-3
	↓	delta-6-desaturase		↓
Gamma-linolenic	18:3n-6			18:4n-3
Dihomogammalinolenic (DGLA)	20:3n-6			20:4n-3
	↓	delta-5-desaturase		↓
Arachidonic (AA)	20:4n-6		Eicosapentaenoic (EPA)	20:5n-3
Adrenic	22:4n-6			22:5n-3
	↓	delta-4-desaturase		↓
	22:5n-6		Docosahexaenoic (DHE)	22:6n-3

FIGURE 1. Outline of EFA metabolism.

and flexibility and modulating permeability and the functioning of membrane-bound enzymes and receptors. Apart from this structural role, the EFAs can be converted by lipoxygenase and cyclo-oxygenase enzyme systems to an astonishing variety of hydroxy-acids, PGs, and LTs. The cyclo-oxygenase system makes PGs and the lipoxygenase system the hydroxy-acids and LTs. The hydroxy acids can be formed from any EFA but the PGs and LTs can be made only from EFAs with 20 or more carbon atoms and 3 or more double bonds. PGs and LTs are sometimes known as eicosanoids. To date five series of PGs have been identified, the 1 series from 20:3 n-6 (DGLA), the 2 series from arachidonic acid (20:4 n-6), the homo-2 series from 22:4 n-6, the 3 series from 20:5 n-3 and the 4 series from 22:6 n-3. Arachidonic acid is the EFA present in greatest abundance and most work has therefore been done on it. However concentration is not the only determinant of importance and the other PG series may prove to have key roles.

There is a tendency for modern researchers to forget the structural roles of EFAs and to regard the eicosanoids as all important. This view is wrong. Columbinic acid is a molecule with an EFA-like structure but which cannot be converted to PGs. Columbinic acid can prevent the emergence of the great majority of the features of EFA deficiency when fed to animals on an EFA-deficient diet.[6]

III. STABILITY OF THE EFAS

Because of their polyunsaturation, the EFAs are susceptible to oxygen attack with the formation of peroxidized fatty acids which are highly toxic and which may be involved in damage to DNA and in the metabolic activation of carcinogens.[7-11] Peroxides may be formed from EFAs prior to consumption or within the body itself. Animals have developed complex mechanisms for protecting themselves against peroxides, both by reducing formation and increasing removal. Antioxidant systems, in which both vitamin E and vitamin C may play important roles, reduce the rate of peroxide production.[11-14] Glutathione, and the selenium-containing enzyme, glutathione peroxidase, are involved in the removal of existing peroxides.[14] There is a quantitative relationship between EFA intake and the level of antioxidant intake required; the higher the amount of PUFA consumed, the greater the need for antioxidant mechanisms.

Unsaturated fats are toxic in the absence of vitamin E[12,13] and a mechanism other than peroxide formation may contribute to this. Vitamin E is able to inhibit the lipoxygenase enzyme system important in formation of hydroxy-acids and LTs.[15] Without adequate vitamin E, large amounts of potentially toxic LTs and hydroxy-acids may be formed.

IV. DIETARY EFAS AND PG FORMATION

There is confusion among researchers concerning the relationship between the amount of linoleic acid in the diet and the rate of PG formation. Many papers seem based on the assumption that there is a direct and positive relationship between the amount of linoleic acid and the amount of PGs formed. This is unjustified. The total absence of EFAs from the diet does indeed lead to low levels of PG formation, but it is not true to say that the higher the linoleic acid intake the more PGs will be formed. There is strong evidence from both humans and animals that if one starts with a normal EFA intake, then a reduction in EFA consumption is consistently associated with a rise and not with a fall in PG production.[3] Studies of the effects of increasing amounts of dietary linoleate on the rate of PG synthesis have shown a multiphasic pattern with no consistent relationship between the two variables.[16,17]

V. INTERPRETATION OF THE CONSEQUENCES OF FEEDING POLYUNSATURATES

It is unfortunate that many authors of papers on EFA or PUFA supplementation in relation to cancer seem to be unaware of the following pitfalls:

1. Care must be taken to ensure that peroxides are not formed prior to administering the EFAs. Food must be made up freshly and/or stored under nitrogen if the polyunsaturated material is not be contaminated with peroxides.

2. The antioxidant content of the oils and diets used should be specified, especially with regard to vitamin E and selenium. Off the supermarket shelf oils which will usually contain synthetic antioxidants such as butylated hydroxy-anisole or hydroxy-toluene (BHA or BHT) are frequently used with comment. Stripped oils which have been deprived of vitamin E are also often used.

3. The antioxidant content of the diet should be proportional to the intake of PUFA. For instance it is common practice to feed two dramatically different PUFA levels (e.g., 0.5 and 20% of calorie intake) without changing the antioxidant intake proportionately. This is likely to give uninterpretable results.

4. *Trans* acids, while they may be chemically polyunsaturates, have no EFA activity and may inhibit EFA metabolism. Many commercially available oils contain substantial amounts of *trans* acids, particularly elaidic, and the levels should be specified.

5. n-3 and n-6 EFAs probably are both important. They certainly each interfere with the other's metabolism. The ratio of the two types of EFA in the diet is therefore probably important. Many studies on EFAs and cancer have given huge amounts of n-6 EFAs with few if any n-3 EFAs probably leading to possibly unrealistic fatty acid compositions in the animals fed in this way.

Virtually none of the papers which will be reviewed is satisfactory in all five respects, and some are unsatisfactory in all five. That this is not simply a trivial theoretical objection is shown by the facts that vitamin E deficiency may enhance and antioxidants may inhibit the development of carcinogen induced tumors in rats.[18,19] Similarly selenium may inhibit mammary tumor growth.[20] Animals deficient in vitamin E and selenium which are then fed PUFAs may become severely immunodepressed.[20] It is therefore difficult to interpret the results of EFA supplementation studies without taking note of these other factors.

VI. HUMAN STUDIES

The human studies are of three types: (1) epidemiological observations on the relation between fat intakes and breast disease, (2) biochemical and endocrine observations on populations at different levels of breast disease risk, and (3) intervention studies in which changes in fat intake or metabolism are examined for their impact on breast cancer incidence or progression. (see Chapter 9). None of the human studies has specifically addressed the issue of EFAs and breast cancer and information about this was for the most part collected incidentally.

A. Epidemiology

There is no doubt that in countries in which total fat consumption is high, breast cancer is common and in ones in which fat consumption is low, breast cancer is less common.[22-31] This question is reviewed in detail in Chapter 5. There is a strong and positive correlation between total fat "disappearance" and breast cancer mortality. There is a less tight but still strong correlation between animal fat and breast cancer. There is no correlation between country by country "disappearance" of vegetable fat and breast cancer (see Carroll[2,32-37]). The great bulk of EFA intake is as vegetable fat. This suggests that variations in EFA intake make little contribution to the fat-related differences in breast cancer mortality.

Migration studies reinforce this conclusion. Japanese in Japan have a low rate of breast cancer.[28,32,38] On migration to the U.S., breast cancer mortality rises, not in the immigrants but in the first generation born in the new country. Total intake of linoleic acid appears similar in the U.S. and Japan, but the ratio of linoleic acid to other dietary fats appears to be much higher in Japan and as a result adipose tissue linoleic acid has been reported to be as much as 50% higher in Japanese in Japan than in Americans.[39] Moreover, total EFA intake is very much higher in Japan than in the U.S. because of the high level of n-3 EFA consumption in the form of eicosapentaenoic acid (EPA) and docosahexaenoic acid (DHE) in seafood.[40] The between-country comparisons therefore suggest that n-6 EFA intake is neutral with regard to human breast cancer and that total EFA intake and/or tissue levels may be negatively correlated with breast cancer.

Within country comparisons are more difficult to perform and to interpret. In Japan breast cancer mortality has recently risen sharply, in close correlation with rises in saturated fat and meat intake.[41] Comparisons of different regions in the U.S. suggest a positive correlation with dairy products, rich both in saturated fats, unusual proteins, and other materials.[42] On the other hand, two studies paradoxically found a negative correlation between breast cancer and egg comsumption.[42,43] Case/control studies involving individual recall of diet have many problems. There is some evidence that dietary total fat and saturated fat but not linoleic acid intake are higher in breast cancer patients than controls[44,45] (see Chapter 5 for a more detailed evaluation).

In another study, beef and pork consumption and use of butter rather than margarine at table and in cooking were higher in breast cancer patients than controls.[43] An important feature of this study, since it accords with the Japanese experience, is that a high consumption of fish was associated with a reduced risk of breast cancer. Another study, published 17 to 25 years after the interviews were conducted, found no difference between breast cancer patients and controls in total fat, animal fat, or vegetable fat consumption.[46]

The most controversial epidemiological observations are those of Enig et al.[4,5] on breast cancer and food consumption trends in the U.S. since the 1920s. Over this period breast cancer mortality has risen but there has been little change in total fat consumption. Dairy and animal fats have been progressively replaced by processed vegetable oils containing *trans* fatty acids. A strong correlation between breast cancer and *trans* PUFA consumption was demonstrated and it was suggested that this might be causal. This view has been strongly challenged, largely on the grounds that feeding of heated and processed vegetable oils to animals on a lifetime basis has no adverse effects of any importance.[47,48] However, although heat-treated oils did not change overall tumor incidence, they did increase breast tumors and reduce those at other sites.[47] *Trans* acids can interfere with the desaturation of linoleic acid.[49] Linoelaidic acid has a clear effect, but elaidic acid which is more frequently consumed has a weak effect. This relatively weak effect may be unimportant in rodents in which desaturation is highly active, but may be significant in humans in whom desaturation is less active.[3] If this is so, the safety of *trans* acids in rats may be irrelevant to the issue of whether they are harmful in humans. Alcohol is also an inhibitor of desaturation[50] and in one unconfirmed report high consumption is associated with an increased breast cancer risk.[51]

B. Biochemical Observations

To the author's knowledge there are no reports of blood EFA levels in women with breast cancer. Total blood lipids may be elevated[45] while cholesterol levels have been reported to be either elevated[52] or normal.[53] The paucity of information is astonishing in view of all that has been written in this field.

Women with benign breast disease have an elevated rate of cutaneous sebum production as compared to normal controls.[54] Women with breast cancer have sebum production rates significantly higher than those in women with benign disease. This may possibly relate to a physiological deficiency of EFAs in women with breast disease because the sebaceous glands hypertrophy under conditions of EFA deficiency.[55]

A major problem in studying the biochemistry of women with breast cancer is that the emotional impact of diagnosis and the nature of the treatment are likely to have profound effects on biochemical parameters. Yet it is the biochemistry prior to these traumatic events which we would like to investigate. Studies on women with benign breast lumps may prove valuable because there is evidence that these women have an increased risk of developing breast cancer.[56-59] A recent study has compared plasma phospholipid fatty acids in controls with two different populations of women with benign breast disease.[60] Saturated fatty acid levels were significantly elevated and linoleic acid concentrations were normal or slightly elevated in the breast disease patients. In spite of the normal linoleic acid, concentrations of linoleic acid metabolites were significantly below normal. Levels of all n-3 EFAs tended to be low. Women with benign breast disease may therefore have a reduced ability to metabolize linoleic acid and possibly a reduced intake of n-3 EFAs.

In contrast to the sparse information of EFAs, there are many studies of PGs and cancer[61] and benign breast disease.[62-66] Women with benign breast disease had significantly higher plasma PGE, present in both phases of the menstrual cycle. Breast tissue from women with fibrocystic disease produced three to four times as much PG as normal during incubation in vitro. This high rate of PG production in benign breast disease is associated with a low level of the 2 series PG precursor, arachidonic acid in plasma phospholipids.[60] This may mean a high rate of arachidonate consumption, or alternatively may be an example of the paradoxical fact that low levels of EFAs are often associated with high rates of PG production.[3]

Breast cancer tissue has been repeatedly shown to produce more 2 series PGs than normal tissue when incubated with or without arachidonic acid.[63,67-74] There is great variability in the results and correlations between PGs and any aspect of tumor biology are rather weak. Estrogen receptor positive cancers may form more PGE than other cancers.[75,76] While undifferentiated lesions made more PGs than differentiated ones, there was unexpectedly an inverse correlation between PG production and tumor size.[64,65,77] There are suggestions that metastatic cancers, especially those which go to bone, may make more PGs than others and that survival time may be reduced when PG production is very high.[64,65,68,69,77-79] However, all associations were very weak and there were many individual exceptions to these generalizations. There is some evidence that hypercalcemia and bone erosion are associated with cancers producing large amounts of PGs and that PGs can erode bone, but there are many patients in whom elevated calcium cannot be accounted for by a PG mechanism.[80-86] There have been few measurements of plasma PG concentrations in breast cancer patients: they have been reported to be either elevated or normal.[87,88]

C. Intervention Studies

The previous section indicates that PG production by breast cancer tissue is definitely higher than that by normal tissue. The data give no indication as to whether this abnormality is simply a secondary consequence of malignant change or whether it may have a role in stimulating the initiation or growth of cancer. Intervention by inhibiting PG synthesis should help to decide this question. Inhibition of PG synthesis failed to reduce plasma calcium or hydroxyproline excretion (a marker of collagen destruction) in women with breast cancer.[89] Nor did inhibition have any effect on time to relapse, occurrence of bony metastases, or survival.[90] These negative results are not conclusive because the drugs may have been unable to produce an adequate suppression of tumor PG formation and "homing" of the drug to the tumor may be required. Tamoxifen is an effective drug in breast cancer which localizes in breast tissue. It is assumed to be an anti-estrogen but it is also an effective PG synthesis inhibitor, an action which could contribute to its clinical efficacy.[91]

There appear to be no studies to date in which changes in fat intake were made with the deliberate intention of modulating the course of breast cancer (see Chapter 9). Feeding linoleic acid-rich diets to normal women caused accumulation of pigments in breast fluids which may have been peroxide related although no peroxide measurements were carried out.[92] Administration of evening primrose oil containing linoleic acid and its metabolite 18:3 n-6 to women with benign breast disease led to significant improvement in women with menstrually related symptoms but not in women with menstrually unrelated symptoms.[93,94] This treatment may have helped to correct the deficit of n-6 metabolites in these women.[60] Administration of vitamin E to women with benign breast disease led to clinical improvement possibly related to reduction in peroxides or hydroxy-acids.[95,96]

High intakes of PUFAs have been taken for many years by individuals, mostly male, at risk of developing coronary heart disease. In a group of such men in Los Angeles assigned to receive a high PUFA diet, there was an increased frequency of colon cancer.[97] Like breast cancer, colon cancer is closely linked to a high total fat intake. However, in five other studies of PUFA-rich diets there was no similar increase in cancer frequency.[98] Reanalysis of the Los Angeles data showed that the increased frequency of colon cancer was in men who were assigned to the high PUFA group but who did not, in fact, follow the diet.[99] A PUFA-rich diet could therefore not be blamed.

D. Conclusions — Human Data

There is strong evidence that a high total fat intake is associated with a high risk of breast cancer. There is no evidence of an increased PUFA intake in patients or populations at high risk. The findings in Japanese women, on sebum production, and on plasma EFAs in benign breast disease suggest that breast cancer may be associated with reduced tissue EFA levels and a low ratio of polyunsaturates to other fats in the diet. There is no doubt that breast cancer tissue makes increased amounts of PGs and this is often naively assumed to indicate excessive EFA intake. However, the relation between dietary EFAs and tissue PG production is complex and there is good evidence that reduced intake of PUFAs may actually increase PG formation. Within the range of EFA levels in human diets there is therefore no evidence that a high EFA intake is positively correlated with breast cancer and there is some tentative evidence of a negative correlation. However all human diets probably contain more than the minimun 3% or so of calories as EFAs which is generally regarded as being necessary for health.[100] The apparently contradictory results which, as will be seen, have been obtained in animal studies almost all relate to EFA intakes below this minimum 3%.

VII. ANIMAL STUDIES

Although ostensibly we study animal cancers with the aim of understanding human cancers, we shall know whether or not the animal cancer is truly a model of the human disease only when we fully understand both. There are strong reasons for questioning whether currently used animal models of breast cancer are indeed relevant to humans. These doubts are particularly important with regard to the question of EFAs and breast disease. This is because EFA metabolism in rats and mice and in humans is dramatically different. There are two rate-limiting steps in the conversion of dietary linoleic acid to arachidonic acid, the precursor of 2 series PGs. The Δ-6-desaturase (D6D) converts linoleic acid to gamma-linoleic acid (GLA, 18:3 n-6) while the Δ-5-desaturase (D5D) converts dihomogammalinolenic acid (DGLA, 20:3 n-6) to arachidonic acid. In rats and mice, both enzymes seem to be highly active and there is rapid conversion of linoleic acid to arachidonic acid.[3,101,102] In adult humans, in contrast, the activity of both desaturases seems to be much lower and there is only restricted conversion of dietary linoleic acid to arachidonic aicd.[3,101-103] Rabbits and guinea pigs are much closer to humans in this respect and might make much better animal models than rats and mice for breast cancer studies.[102] Morever, in humans the ratio of n-3/n-6 EFAs is considerably higher than in rats and mice, suggesting that considerable caution must be invoked in interpreting experiments which involve dramatic changes from normal in the n-3/n-6 ratio such as will inevitably happen when animals on a basic fat-free diet are supplemented with large amounts of safflower oil or corn oil.[102] For these reasons there must be serious reservations about the validity of experiments on rats and mice as far as human breast cancer and EFAs are concerned. Unfortunately, guinea pigs and rabbits are expensive and little is known about mammary tumors in these species. But these are not valid reasons for the near-exclusive use in EFA studies of animals whose EFA metabolism is completely different from that in humans.

It is important to remember that EFAs are essential nutrients. They play a major role in the normal development of the mammary gland and are constituents of all cell membranes.[104-107] It is not unlikely, therefore, that malignant tumors derived from mammary glands require EFAs for growth. As with any essential nutrient, it will, therefore, be possible to show that deprivation of EFAs will inhibit tumor growth and that provision of the nutrient will stimulate growth again. But such inhibition can be achieved only at the cost of producing an EFA deficiency state. This may be tolerable in the short term, especially since tumor requirement for essential nutrients may be greater than those of their normal parent tissues. It may therefore be possible to find a level of essential nutrient intake which limits tumor growth but at least in the short term has little or no effect on the normal tissues of the body. This may prove a legitimate practical strategy in short-term attempts to limit the growth of an existing tumor in someone whose life expectancy is limited. It is not a legitimate strategy for the nutritional prevention of cancer or for treatment when the life expectancy of the individual is considerable.

This means that experiments on animal cancers in which the daily EFA intake is below that regarded as essential in humans are of strictly limited value. Animal studies based on growth and biochemical analysis of tissues suggest that 1 to 2% of total calories in the form of linoleic acid is the minimum required to prevent EFA deficiency.[108,109] Mammary development in rodents may fail at EFA levels which maintain normal growth, suggesting that more linoleic acid, perhaps 2 to 3% of total calorie intake, may be required for the mammary gland.[107] A WHO Expert Committee which reviewed EFA requirements in humans suggested that a minimum of 3% of total calories should be provided as linoleic acid, rising up to 4 to 5% in pregnant and lactating women and children.[100] Almost all populations in the world, including those with very low rates of breast cancer, take at least 3% of dietary calories in the form of linoleic acid.[32-36] In view of the probable importance of EFAs in reducing the risk of cardiovascular disease, I do not believe that anyone would seriously suggest that linoleic acid intake should be reduced below 3% of total calories. The important question for humans which the animal studies should address is, therefore, whether increasing EFA intake above this level is likely to have any deleterious effect on the incidence or progression of breast cancer.

A. Effects of EFAs on Animal Mammary Cancers

The studies of Tannenbaum, Silverstone, and Benson were the pioneering efforts which showed that a diet exceptionally rich in fats could enhance the developments of various types of animal tumors.[1,110,111] Carroll's group reopened the issue of fat and animal cancer and they have produced an impressive series of studies, for the most part on DMBA-induced tumors in rats.[2,32-37,112-118] Various levels of saturated, monounsaturated, and polyunsaturated fats were administered before and after the administration of DMBA. A detailed historical review of this topic is found in the chapter by Cohen. The main conclusions to be drawn from this long series of studies are as follows:

1.   Animals on a high-fat diet develop more tumors earlier than animals on a low-fat diet.
2.   Animals on an EFA-deficient diet have a reduced risk of tumor formation.
3.   The fat effect is not related to any action on the absorption of lipid-soluble carcinogen (DMA) or on its uptake into mammary tissue.
4.   A certain minimum level of EFA intake is required for maximum tumor development. About 2 to 3% of total calorie intake as EFAs seems to be the critical figure. Provided that this is present in the diet, high intakes of any type of fat, saturated, monounsaturated, and polyunsaturated, are effective in further increasing tumor growth. Either n-6 or n-3 EFAs seem able to fulfill the minimum requirement. Since n-3 EFAs inhibit PG formation from arachidonic acid, this argues against PG formation being critical.
5.   Administering a high-fat intake prior to the carcinogen with a normal or low-fat intake afterwards seems to have no effect on tumor growth. In contrast, even many weeks after giving DMBA, adding fat to or withdrawing it from the diet is able to modulate tumor growth, although this is much less clear cut. The question is reviewed in detail in Chapter 6. This has been interpreted as indicating that the fat effect is not on the initiation of tumors but on their promotion and subsequent growth.
6.   Rapeseed oil is an exception to the general rule that PUFA-rich oils stimulate tumor growth. Rapeseed oil contains erucic acid which has two possibly relevant actions. It inhibits the conversion of linoleic acid to arachidonic acid and it inhibits the lipoxygenase enzyme which is involved in the conversion of PUFAs to hydroxy-acids and leukotrienes.[119]

Other groups have performed studies which on the whole confirm and amplify the findings of Carroll et al. Ip et al. compared the effects of diets containing 0.5% or 20% of corn oil, given before or after DMBA, on tumor yield.[120,121] The high-fat diet was highly effective in increasing tumor yield even when it was started as late as 20 weeks after DMBA at 50 days. It was also effective even when DMBA was given at 150 days, a time which in animals on a normal diet usually has little effect on tumor development. Chan et al. emphasized that most of the high saturated fat diets used were probably marginally deficient in EFAs and that the reduced rates of tumor development on these diets as compared to PUFA-rich diets were probably simply due to lack of an essential nutrient.[122] On the other hand, when they looked at the effects of a variety of diets on NMU-induced tumors they did find a good correlation between tumor yield and linoleic acid intake and an even stronger one with the combined intakes of linoleic and oleic acids. There was no correlation with oleic acid alone. Dao et al. reopened the issue of initiation vs. promotion.[123] They noted that high-fat diets before carcinogen exposure were always given for much shorter intervals than the same diets after carcinogen exposure. If there was prolonged exposure to a high-fat diet prior to giving the carcinogen then there did appear to be an effect on tumor growth.

Dayton et al. attempted to sort out the importance of linoleic acid by administering two types of safflower oil at 20% of total calories.[124] One contained 13% of linoleic acid and the other contained 70% of linoleic acid. There was no difference in DMBA-induced tumor growth in the two rat groups. This agrees with Carroll's idea that once a threshold of 2 to 3% of dietary calories has been exceeded, linoleic acid is no more effective than any other fat in promoting tumor development.

Only a few studies have been performed on tumor and normal mammary gland lipids as influenced by carcinogens. Different rat strains differ dramatically in their susceptibility to carcinogenesis but all had similar linoleic acid levels in mammary tissue.[125] Linoleic acid metabolites were not measured and this would be worth doing. A major reason for this, of course, is that normal mammary tissue is mainly adipose, while tumors contain many more epithelial cells. Linoleic acid was reduced in the tumor tissue but arachidonic acid was normal.[126] Another group found reduced linoleic acid in DMBA tumor phospholipids with normal arachidonic acid: arachidonic acid was elevated in tumor triglycerides.[127]

In human cancers, some forms of treatment may reduce tumor size but have no effect on patient survival. Goodman et al. reviewed the tumor patterns and longevity of Osborne-Mendel rats fed corn oil-supplemented chow or normal chow, both groups being used as controls in programs of carcinogen testing.[208] The corn oil-fed animals had rather more mammary tumors than the chow-fed ones, but the former actually lived longer.

The above rat studies have been complemented by broadly similar ones in mice. Using a variety of different transplantable mammary tumors, 15% corn or safflower oil in the diet was found to produce much more rapid tumor growth than a fat-free diet.[128-129] With some tumors as little as 0.1% linoleic acid added to the diet had as much effect as 15% corn oil but other tumors required more linoleic acid. When fat-free diets were supplemented with oleic acid (18:1 n-9), linoleic, α-linolenic (18:3 n-3), or columbinic acid, only the linoleic group showed increased tumor growth. This suggests that in contrast to rat DMBA tumors, these mouse transplantable tumors do not react to n-3EFAs. The problems inherent in interpreting the results of studies using transplantable tumors are discussed by Cohen in Chapter 6.

Mice either treated with DMBA or given transplantable mammary tumors, showed stimulation of tumor growth when a high-fat diet was given afterwards but not before. These observations led Hopkins et al. to propose that the main effect of fat was on tumor promotion and growth and not on initiation,[130-132] a conclusion which had earlier also been drawn by Carroll.[115]

Some studies have obtained results not in accord with the apparently clear picture described so far. Brown emphasized that most studies have not clearly defined the compositions or antioxidant properties of diets and have fed diets bearing little relation to anything likely in humans.[133] He compared animals fed 5 or 17% of calories as fat and used saturated fats, oleic acid, linoleic acid, and *trans* fatty acids with varying levels of BHA as an antioxidant. This careful investigation could demonstrate only weak and inconsistent differences between the effects of the various diets on spontaneous and dimethylhydrazine-induced tumors. This raises the possibility that realistic within-population individual differences in type and amount of fat intake may have little effect on human breast cancers. Gridley et al. in an even more complex experiment tested 14 different diets with different fats being fed at 5 or 30% of calorie intake and different proteins at 11 or 33% on the development of spontaneous mammary tumors.[134] Lowest tumor frequencies were in the low-fat, low protein groups, but a high corn oil intake had only a small and nonsignificant effect. Moreover, survival to 120 weeks was considerably greater in the high corn oil group than the high butter group, arguing against any negative effect of EFAs on survival. A further degree of sophistication was reached by Tinsley at al., who looked at spontaneous mammary tumors in mice fed diets containing 10% of 11 different fats and oils and 9 mixtures thereof.[135] They used multiple regression analysis to assess the contribution of individual fatty acids, pointing out the shortcomings of simple studies in which an increased intake on one nutrient inevitably means a reduced intake of another making it difficult to know which is responsible for the effect. Oleic acid and 18:3 n-3 had no effect, again casting doubt on the importance of n-3 EFAs. Erucic and stearic acids both appeared to be inhibitory to tumor development while linoleic acid had a modest stimulating effect.

Transplantable tumors in rats have been much less studied than in mice. Growth of the R3230AC tumor was inhibited in animals on a fat-free diet and stimulated by very small doses of linoleic acid.[129] In contrast, relatively small doses of evening primrose oil, containing both linoleic acid and its metabolite, gamma-linolenic acid (18:3 n-6) inhibited growth of this same tumor line.[136] This suggests that gamma-linolenic acid and linoleic acid may have quite different effects. In a number of animal and human cancers, there are strong indications that the ability to convert linoleic acid to gamma-linolenic acid has been lost.[137] This would lead to loss of the ability to make PGE₁, an important stimulator of cyclic AMP formation. It was proposed that 18:3 n-6 might cause some or all of the properties of cancer cells to revert to normal. Gamma-linolenic acid has been found to have potent inhibitory effects on the growth of a variety of human and animal malignant cell lines without blocking the growth and division of normal cells.[138-140] These studies suggest that much more attention must be paid to the actions of individual members of the n-6 EFA series. It is not sufficient to administer linoleic acid and to assume either that it will be converted to the other n-6 EFAs or that the other n-6 EFAs will have similar effects.

VIII. MECHANISM OF ACTION OF EFAs

It is clear that restriction of linoleic acid intake to levels below the minimum required for normal EFA function limits the growth of mammary tumors in animals. It is also clear that, once the EFA requirement has been fulfilled, provision of fat at high levels stimulates mammary tumor growth. It is less clear whether the type of fat matters once the EFA requirement has been satisfied. The balance of evidence from the animal studies suggests either that it does not or that if EFAs do have any greater effect than other fats the difference is rather small.

It seems to me likely that the mechanisms of growth restriction of suboptimal EFA intakes, and mechanisms of growth stimulation by fat once optimum EFA intake has been reached will prove to be quite different. This important distinction has rarely been considered when mechanisms have been discussed in the past. Unfortunately many of the experimental designs (e.g., comparisons of 0.5 and 15% corn oil added to a fat-free diet) preclude any clear separation between the two situations. In future studies on mechanisms should pay more attention to this point.

The major mechanisms which have been considered to be involved in the effects of fats and cancer will be briefly reviewed. They have also been reviewed by Carroll's group, Dao and Chan,[123] and Hopkins and West.[130]

A. Essential Fatty Acids and Prostaglandins

A distinction must be made between direct effects of EFAs and derived products like PGs on mammary cells themselves, and indirect effects mediated through changes in the mammary environment, such as modifications of hormone secretion or immune function. Both direct and indirect effects may well play a role (see Chapter 6).

Direct effects can be most readily demonstrated in cell or organ culture. Cell lines derived from DMBA tumors failed to grow in the absence of linoleic acid: once minimal growth requirements had been fulfilled, increasing concentrations of linoleic acid continued to enhance proliferation.[141] Arachidonic acid had little or no effect on cell multiplication, with oleic and α-linolenic having intermediate actions.[142] It is impossible to generalize about PG effects. PGF$_{2\alpha}$ was a growth stimulating factor for the MCF-7 human breast cancer cell line.[143] PGE₁, in contrast, is a potent differentiating agent. Cells derived from DMBA tumors and exposed to PGE₁ in vitro were substantially less effective at generating tumors when injected into nude mice.[144] PGE₂ and PGF$_{2\alpha}$ had similar but less potent actions. On the other hand, a whole range of PGs, including PGE₁, were able to increase the production of mouse mammary tumor viruses.[145] At present any conclusions which can be drawn from these direct experiments are very limited and much more attention must be paid to the testing of specific EFAs and PGs at physiologically relevant concentrations.

The dependence of normal mammary gland differentiation on dietary EFAs has been well established. Addition of cholesterol to the diet accentuated the effects of EFA deficiency,[106] on mammary gland development, suggesting that a linoleic acid metabolite may be important because dietary cholesterol has been shown to inhibit linoleic desaturation in vivo.[146] Indomethacin also has some inhibitory effect on normal mammary gland development indicating that PGs may be important since the drug blocks EFA conversion to PGs by the cyclo-oxygenase enzyme.[147]

Indirect ovarian effects may be important in the effects of EFA deficiency on the mammary glands. Ovarian function was defective in these animals and there were no corpora lutea.[104-106]

Many attempts have been made to correlate the biological effects of the diets with changes in fatty acid composition of normal and tumorous mammary tissue. In mice a fat-free diet lowered the overall linoleic acid level in transplantable tumors and halved the arachidonate content of phospholipids but not of triglycerides.[148,149] 1% corn oil was as effective as 15% corn oil both in stimulating tumor growth and in raising the tumor arachidonate content, although the 15% diet was much more potent in raising linoleic acid levels. The fact that as little as 0.1% dietary linoleic acid could stimulate tumor growth, even though the tumors in animals on a fat-free diet contained reduced but still substantial amounts of linoleic and arachidonic acids, is important. It suggests that much of the EFA content of the mammary gland is locked in fractions which have little relevance to tumor growth, and that dietary linoleic acid rapidly changes some unknown fraction which is relevant. Unexpectedly, although these studies suggested that arachidonic acid might be the important EFA, dietary arachidonate was much less effective than linoleic acid in enhancing the growth of tumor implants.[129] When animals with transplanted tumors were fed corn oil, arachidonic acid concentrations rose least in the most rapidly growing tumors and there was little correlation between growth and arachidonic acid levels.[150,151] In contrast to these studies, in NMU-induced tumors a high EFA diet stimulated tumor growth, raised linoleic acid levels in tumors, but actually lowered tumor and liver arachidonic acid concentrations.[152] Also using NMU-induced tumors, Chan et al. found that of five groups of rats only the one fed a high corn oil diet had substantial amounts of arachidonic acid in the normal gland, while in all groups there were reduced linoleic and elevated arachidonic concentrations in the tumor phospholipids as compared to the normal gland. Obviously changes in fatty acid compositions are complex and difficult to interpret and there are only weak correlations between the fatty acid compositions of diets and tumors.

In contrast, there does seem reasonable evidence that the effects of EFA supplementation within the range of 0 to 3% of total calories are in part dependent on PG synthesis. There is virtually no evidence concerning the roles of PGs in the tumor stimulation by fat which occurs after minimum EFA requirements have been met. PGs seem to play an important role in various aspects of mammary physiology, including the formation of lactalbumin[153] and the mechanism of action of prolactin.[154-156] DMBA-induced tumors synthesize significantly more PGs than normal glands with a greater rise in PGE_2 than in $PGF_{2\alpha}$.[157] In a range of mouse mammary tumors of varying metastatic potential there was a positive correlation between PGE level and metastatic activity.[158] DMBA-induced tumors contained more arachidonic acid and synthesized more PGs than normal glands. On the other hand, in spontaneous tumors in mice, while PGF levels were elevated, PGE concentrations were actually reduced.[160] Two groups found an inverse relationship between tumor PGE and metastatic activity: tumor regression occurring with ovariectomy was actually associated with a sharp rise in tumor PGE levels.[161,162] Hormone-dependent tumors bound PGE_1 specifically whereas loss of hormone-dependence was associated with loss of PG binding.[163] In NMU-induced tumors in rats there was elevated PGE_2, $PGF_{2\alpha}$, 6-keto-PGF_1, and thromboxane B_2 in the tumors with much greater rises in the first two than the last two when arachidonic acid was added.[164] There was an inverse correlation between tumor size and PG content. It therefore seems that while tumors are almost invariably associated with increase PG formation, PGs may be involved in feedback mechanisms tending to limit tumor growth. PGEs have, of course, been found effective in inducing reverse transformation in a variety of human and animal transformed cell lines.[165,166]

One way of investigating the role of PGs is to administer agents which will inhibit PG synthesis. The weak inhibitor, aspirin, had no effect on growth of transplantable tumors in mice, but the more potent indomethacin and tetraynoic acid (TYA) were effective in limiting growth: TYA also inhibited conversion of linoleic acid to arachidonic.[128,129,167] Another group also found that indomethacin inhibited transplanted mouse mammary tumor growth,[168,207] while flurbiprofen, another PG synthesis inhibitor, reduced the tumorigenic effect of NMU.[169] Indomethacin completely abolished the enhancing effect on DMBA tumors of 18% as opposed to 5% corn oil, providing a suggestion that PGs may be involved in the further stimulation of tumor growth which occurs when EFA intake is fully adequate.[170,171] Curiously, when indomethacin was given before DMBA there appeared to be an increased tumor yield early on followed by increased regression later, suggesting that PGs may inhibit tumor initiation, but stimulate growth of established tumors. There have also been hints that a high EFA intake before giving DMBA may actually inhibit tumor growth.[121] These paradoxical hints should not be ignored and deserve full investigation for they have major potential implications for human cancer.

Apart from the results on reverse transformation, there is evidence that PGs can be oncolytic and this should not be ignored. The oncolytic effects appear to be related to inhibition of ovarian function.[172,173]

Bennett's group, in addition to their work on human breast cancer, have also provided substantial evidence that transplantable mammary tumors in mice treated with combinations of flurbiprofen, surgery, and chemotherapy do better than those treated without PG synthesis inhibition.[174,175]

In summary, there is modestly convincing evidence that EFAs and PGs are necessary for the growth of both normal mammary glands and mammary tumors and that EFA and PG levels which are inadequate to sustain normal gland development will not sustain tumor growth. The lack of effect of arachidonate on tumor growth is puzzling and it is important to note that linoleic acid may both stimulate and inhibit PG synthesis depending on the conditions without being converted to arachidonic acid by modulating the availability of the PG precursor.[3,176] The antitumor effects of PGs and of EFAs such as 18:3 n-6 must not be ignored and much work needs to be done before an integrated picture can emerge.

B. Prolactin

Most carcinogen-initiated tumors in rodents are stimulated by prolactin. Interestingly, prolactin given before a carcinogen consistently inhibits tumor growth and there is one well-known rat tumor, R3230AC, which is inhibited by prolactin. Unfortunately in these respects animal and human tumors seem very different and attempts to modify human breast cancer by regulating prolactin have proved disappointing.[177,178]

Chan et al. were the first to draw attention to the possibility that a high-fat diet might act by enhancing prolactin secretion[178-183] (see Chapter 6). Ovariectomy or anti-estrogens inhibited carcinogen-induced tumor development in both high- and low-fat groups but the differential between the two still persisted. In contrast, bromocriptine which inhibits prolactin secretion, completely abolished the high-fat/low-fat differential. Female animals on a high-fat diet showed elevated ether-stimulated prolactin levels at proestrus. Ip et al. also found elevated proestrous prolactin levels in animals on a high EFA diet.[184] In an experiment the reverse of giving bromocriptine, they then made animals with median eminence lesions which produced chronically high prolactin levels. In this situation the high-fat diet still produced a higher tumor yield than the low-fat diet. In haloperidol-treated rats, which also have high prolactin levels, the high-fat diet was still effective.[185]

Thus prolactin seems to be necessary for the fat effect, but changes in prolactin secretion may have only a partial role. Changes in prolactin action, which could occur even though prolactin levels are high and constant, may be equally or more important. Such changes could operate at the receptor or postreceptor levels and there are sound reasons for considering both. EFAs seem required for normal prolactin receptor development and high corn oil diets have been shown to increase prolactin binding in both liver and tumor in rats with NMU-tumors.[104-106,152] The evidence that postreceptor second messenger effects of prolactin may be partially mediated by prostaglandins was reviewed earlier.

Thus there are strong reasons for believing that prolactin effects are very important in the mechanism of EFA stimulation of rodent mammary tumors. In view of the weak evidence for prolactin involvement in human breast cancer,[177] the lessons from these animal models must therefore be interpreted cautiously.

C. Vitamin E and Antioxidants

Polyunsaturated fatty acids (PUFAs) can readily give rise to peroxides and hydroxy-acids either spontaneously given the right conditions or with the assistance of lipoxygenase enzyme systems. Fatty acid peroxides are toxic and carcinogenic.[7,8] Vitamin E has a dual action since it is both an antioxidant and an inhibitor of lipoxygenase enzymes. In spite of the potential importance of this question, only a few studies of EFAs and mammary cancer have considered antioxidants. Dayton et al. found that added vitamin E had no effect on DMBA-induced tumor growth in animals fed high doses of safflower oil.[124] Another group found that while vitamin E and BHA had no effect on DMBA-induced tumors, two other antioxidants, BHT and propyl gallate, did inhibit tumor growth, confirming earlier observations.[19,186-190] Rapeseed oil, which contains erucic acid, has consistently been shown to be much less tumor promoting than might be expected. The importance of the lipoxygenase in EFA effects is suggested by the recent observation that erucic acid is a potent inhibitor of this enzyme.[194]

D. Selenium

Selenium is a constituent of glutathione peroxidase, which is involved in the removal of peroxides. The evidence relating selenium and breast cancer is reviewed elsewhere in this volume by Medina. Certainly selenium deficiency enhances and selenium excess reduces the effects of EFAs in enhancing tumorigenesis in response to DMBA.[192-194]

E. Immunological Abnormalities

The immune system has been thought to play a critical role in the regulation of tumor development. Again this may be more true of animal cancer models than of most human cancers. The cancers which develop in immunosuppressed humans form an unusual group, with lymphoid system malignancies being overwhelmingly important and with little evidence of any effect on the common solid tumors such as breast cancer. There is excellent evidence that increasing the EFA intake in animals can suppress the development of various autoimmune phenomena although it may actually simultaneously enhance the PHA responsiveness of lymphocytes.[195-199] The immunosuppressive effect may be dramatically enhanced if there are simultaneous deficiencies of vitamin E and selenium.[21]

Fats may have two quite distinct effects on the immune response. They might, as suggested by the papers just referred to, have a primary immunosuppressing effect, making it easier for tumors to begin and to grow. On the other hand, evidence reviewed in this paper convincingly shows that tumors make more PGs than normal tissue. PG production by tumors may have an immunosuppressive effect, allowing the tumor to grow more quickly.[200-203] EFAs, by providing more PG precursors to the tumor, could enhance this subversion of the immune system.

Wagner et al. performed a highly original study of the interaction between the immune system and the effects of corn oil supplementation on DMBA-induced tumors.[204] In animals given 0.5 or 5% corn oil, thymectomy at 35 days substantially reduced the carcinogenic effect of DMBA given at 50 days. This protective effect was lost in animals given 20% corn oil. Thymectomy lowered proestrous prolactin but the high corn oil diet elevated prolactin in both thymectomized and sham-thymectomized animals. It was suggested that thymectomy might work partly through prolactin but mainly through enhancing immune activity by preferential inactivation of T suppressor lymphocytes and a consequent rise in helper and natural killer cell activity.

F. Other Possible Mechanisms

This brief review by no means exhausts the possible mechanisms of EFA action on tumor growth.[123,130] Effects on the vascularity of the tumor and its bed, effects on aryl hydrocarbon hydroxylase related carcinogen activation,[205] and effects on sialic acid metabolism[206] are all real possibilities. It seems to me probable that no single mechanism is involved and that the search for a unitary explanation is probably futile. Tumor initiation and promotion are complex processes which may be enhanced or inhibited at many points. Just because a car engine stops when the fuel pump fails does not mean that the fuel pump is the only essential mechanism which keeps the car running. In any mechanism involving a complex chain of events, interruption at any point in the chain will block the mechanism. Too often advocates of a particular mechanism of tumor initiation or promotion adopt the fuel pump argument. Because blockade of one particular mechanism interferes with the whole system, they reason that their mechanism is the only important one. This is patently foolish and full understanding will almost certainly require the integration of many different concepts (see Figure 5 in Chapter 6).

IX. CONCLUSIONS

At first sight the animal and human studies seem to be in conflict particularly with regard to PUFA intake. Most of the human evidence points to total fat and/or saturated fat as important in enhancing breast cancer, exonerating vegetable fats and EFAs, while most of the animal evidence suggests that EFAs are involved. It is possible that mechanisms of mammary carcinogenesis in humans and rodents are completely different and that the animal models are irrelevant to the human disease, partly because of their nature and partly because EFA metabolism is somewhat different in humans and rats and mice. On the other hand, it seems to be probable that some reconciliation will come from development of Carroll's suggestion that below 3% of total calorie intake variations in EFA consumption are vital and that above this level all fats have some tumor-enhancing properties. The effects demonstrable at EFA intakes of below 3% of total calories may be related to subclinical EFA deficiency states. They have little practical importance for human cancer, except for the very limited possibility that in women with terminal breast cancer and a very short life expectancy restriction of EFA intake might conceivably provide some palliation. It may be that for human breast cancer, as for human cardiovascular disease, it is not so much the total EFA intake which matters, but rather the ratio of PUFAs to saturated fats (the so-called P/S ratio). Countries like Japan with a high P/S ratio in the national diet have low rates of breast disease and of heart disease and more attention should be paid to this ratio in studies of human breast cancer. It is important that in an obsession with one disease the fact that people can die of others should not be forgotten. Current life expectancies are higher in

Japan than in any other country which would suggest that for overall health we should aim to reduce total fat intake, with the brunt of the reduction coming in the form of saturated fats leading to a rise in the P/S ratio. Such a change in North America and Europe may be expected to lead to a reduction of mortality from both breast cancer and cardiovascular disease.

REFERENCES

1. **Tannenbaum, A.,** The genesis and growth of tumors. III. Effects of a high fat diet, *Cancer Res.,* 2, 468, 1942.
2. **Carroll, K. K.,** Experimental evidence of dietary factors and hormone dependent cancers, *Cancer Res.,* 35, 3374, 1975.
3. **Horrobin, D. F.,** The regulation of prostaglandin biosyntheses by the manipulation of essential fatty acid metabolsim, *Rev Drug Metab. Drug Interact.,* 4, 339, 1983.
4. **Enig, M, G., Munn, R. J., and Keeney, M.,** Dietary fat and cancer trends — a critique, *Fed. Proc.,* 37, 2215, 1978.
5. **Enig, M. G., Munn, R. J., and Keeney, M.,** Dietary fats and cancer trends — response, *Fed. Proc.,* 38, 2437, 1979.
6. **Houtsmuller, U. M. T.,** Columbinic acid; a new type of essential fatty acid, *Prog. Lipid Res.,* 20, 889, 1982.
7. **Cutler, M. G. and Schneider, R.,** Sensitivity of feeding tests in detecting carcinogenic properties in chemicals: examination of 7,12-dimethylbenz[a]anthracene and oxidized linoleate, *Food Cosmet. Toxicol.,* 11, 443, 1973.
8. **Cutler, M. G. and Schneider, R.,** Tumours and hormonal changes produced in rats by subcutaneous injections of linoleic and hydroperoxide, *Food Cosmet. Toxicol.,* 12, 451, 1974.
9. **Summerfield, F. W. and Tappel, A. L.,** Peroxidative damage to DNA by polyunsaturated fats and protection by vitamin E, *Age,* 6, 144, 1984.
10. **Reigh, D. L., Stuart, M., and Floyd, R. A.,** Activation of the carcinogen N-hydroxy-2-acetylaminofluorene by rat mammary peroxidase, *Experientia,* 34, 107, 1978.
11. **Logani, M. K. and Davies, R. E.,** Lipid oxidation: biologic effects and antioxidants — a review, *Lipids,* 15, 485, 1980.
12. **Horwitt, M. K.,** Status of human requirements for vitamin E, *Am. J. Clin. Nutr.,* 27, 1182, 1974.
13. **Horwitt, M. K.,** Vitamin E: a reexamination, *Am. J. Clin. Nutr.,* 29, 569, 1976.
14. **Hoekstra, W. G.,** Biochemical function of selenium and its relation to vitamin E, *Fed. Proc.,* 34, 2083, 1975.
15. **Panganamala, R. V., Miller, J. S., Gwebu, E. T., Sharma, H. M., and Cornwell, D. G.,** Differential inhibitory effects of vitamin E and other antioxidants on prostaglandin synthetase, platelet aggregation and lipoxidase, *Prostaglandins,* 14, 261, 1977.
16. **Sullivan, L. M. and Mathias, M. M.,** Eicosanoid production in rat blood as effected by fasting and dietary fat, *Prostaglandin Leukotrienes Med.,* 9, 223, 1982.
17. **Dupont, J., Mathias, M. M., and Connally, P. T.,** Effects of dietary essential fatty acid concentration upon prostanoid synthesis in rats, *J. Nutr.,* 110, 1695, 1980.
18. **Lee, C. and Chen, C.,** Enhancement of mammary tumorigenesis in rats by vitamin E deficiency, *Proc. Am. Assoc. Cancer Res.,* 20, 132, 1979.
19. **Harman, D.,** Dimethylbenz(a)anthracene induced cancer: inhibiting effect of dietary vitamin E, *Clin. Res.,* 17, 125, 1969.
20. **Medina, D., Lane, H. W., and Shepherd, F.,** Effect of dietary selenium levels on 7,12-dimethylbenzanthracene-induced mouse mammary tumorigenesis, *Carcinogenesis,* 4, 1159, 1983.
21. **Sheffy, B. E. and Schultz, R. D.,** Nutrition and the immune response, *Cornell Vet.,* 68, 48, 1978.
22. **Doll, R. and Peto, R.,** The causes of cancer: quantitative estimates of available risks of cancer in the United States today, *J. Natl. Cancer Inst.,* 66, 1191, 1981.
23. **Donegan, W. L., Hartz, A. J., and Rimm, A. A.,** The association of body weight with recurrent cancer of the breast, *Cancer,* 41, 1590, 1978.
24. **Hankin, J. H. and Rawlings, V.,** Diet and breast cancer: a review, *Am. J. Clin. Nutr.,* 31, 2005, 1978.
25. **Hems, G.,** The contributions of diet and childbearing to breast cancer rates, *Br. J. Cancer,* 37, 974, 1978.
26. **Kelsey, J. L.,** A review of the epidemiology of human breast cancer, *Epidemiol. Rev.,* 1, 79, 1979.

27. **Kinlen, L. J.,** Fat and cancer, *Br. Med. J.*, 286, 1081, 1983.
28. **Nagasawa, H.,** Nutrition and breast cancer: a survey of experimental and epidemiological evidence, *IRCS J. Med. Sci.*, 8, 786, 1980.
29. Committee on Diet, Nutrition and Cancer, Diet, Nutrition and Cancer, National Academy Press, Washington, D.C., 1982.
30. **Willett, W. C. and MacMahon, B.,** Diet and cancer: an overview. II, *N. Engl. J. Med.*, 310, 697, 1984.
31. **Vorherr, H.,** *Breast Cancer*, Urban and Schwarzenberg, Baltimore, 1982.
32. **Carroll, K. K.,** Essential fatty acids: what level in the diet is most desirable?, *Adv. Exp. Med. Biol.*, 83, 535, 1977a.
33. **Carroll, K. K. and Khor, H. T.,** Effects of dietary fat and dose level of 7,12-dimethylbenz(α)-anthracene on mammary tumor incidence in rats, *Cancer Res.*, 30, 2260, 1970.
34. **Carroll, K. K. and Khor, H. T.,** Effects of level and type of dietary fat on incidence of mammary tumors induced in female Sprague-Dawley rats by 7,12-dimethylbenz(α)anthracene, *Lipids*, 6, 415, 1971.
35. **Carroll, K. K.,** Dietary factors in hormone-dependent cancers, *Curr. Concepts Nutr.*, 6, 25, 1977b.
36. **Carroll, K. K.,** Lipids and carcinogenesis, *J. Environ. Pathol. Toxicol.*, 35, 253, 1980.
37. **Carroll, K. K. and Davidson, M. B.,** The role of lipids in tumorigenesis, in *Nutrition and Cancer*, Arrolt, M. S., van Eys, J., and Wang. Y-M, Eds., Raven Press, New York, 1982, 237.
38. **Buell, P.,** Changing incidence of breast cancer in Japanese-American women, *J. Natl. Cancer Inst.*, 51, 1479, 1973.
39. **Insull, W., Lang, P. D., and Hsi, B. P.,** Studies of arteriosclerosis in Japanese and American men. 1. Comparison of fatty acid composition of adipose tissue, *J. Clin. Invest.*, 48, 1313, 1969.
40. **Kagawa, Y., Nishizawa, M., and Suzuki, M.,** Eicosapolyenoic acids of serum lipids of Japanese islanders with low incidence of cardiovascular disease, *J. Nutr. Sci. Vitaminol.*, 28, 441, 1982.
41. **Hirayama, T.,** Epidemiology of breast cancer with special reference to the role of diet, *Prev. Med.*, 7, 173, 1978.
42. **Gaskill, S. P., McGuire, W. L., Osborne, C. K., and Stern, M. P.,** Breast cancer mortality and diet in the United States, *Cancer Res.*, 39, 3628, 1979.
43. **Lubin, J. H., Burns, P. E., Blot, W. J., Ziegler, R. G., Lees, A. W., and Fraumeni, J. F.,** Dietary factors and breast cancer risk, *Int. J. Cancer*, 28, 685, 1981.
44. **Miller, A. B.,** Role of nutrition in the etiology of breast cancer, *Cancer*, 39, 2704, 1977.
45. **Basu, T. K. and Williams, D. C.,** Plasma and body lipids in patients with carcinoma of the breast, *Oncology*, 31, 172, 1975.
46. **Graham, S., Marshall, J., Mettlin, C., Rxepka, T., Nemoto, T., and Byers, T.,** Diet in the epidemiology of breast cancer, *Am. J. Epidemiol.*, 116, 68, 1982.
47. **Poling, C. E., Eagle, E., Rice, E. E., Durand, A. M. A., and Fisher, M.,** Long term responses of rats to heat-treated dietary fats. IV. Weight gains, food and energy efficiencies, longevity and histopathology, *Lipids*, 5, 128, 1969.
48. **Applewhite, T. H.,** Statistical correlations relating trans fats to cancer: a commentary, *Fed. Proc.*, 38, 2435, 1979.
49. **Hill, E. G., Johnson, S. B., Lawson, L. D., Mahfouz, M. M., and Holman, R. T.,** Perturbation of the metabolism of essential fatty acids by dietary partially hydrogenated vegetable oil, *Proc. Natl. Acad. Sci. U.S.A.*, 79, 953, 1982.
50. **Nervi, A. M., Peluffo, R. O., and Brenner, R. R.,** Effect of ethanol administration of fatty acid desaturation, *Lipids*, 15, 263, 1980.
51. **Rosenberg, L., Slone, D., Shapiro, S., et al.,** Breast cancer and alcoholic beverage consumption, *Lancet*, 1, 267, 1982.
52. **Dickerson, J. W.,** Nutrition and breast cancer, *J. Hum. Nutr.*, 33, 17, 1979.
53. **Hiatt, R. A., Friedman, G. D., Bawol, R. D., and Ury, H. K.,** Breast cancer and serum cholesterol, *J. Natl. Cancer Inst.*, 68, 885, 1982.
54. **Goolamali, S. K. and Shuster, S.,** A sebotrophic stimulus in benign and malignant breast disease, *Lancet*, 1, 428, 1975.
55. **Sinclair, H. M.,** Essential fatty acids and the skin, *Br. Med. Bull.*, 14, 258, 1958.
56. **Coombs, L. J., Lilienfeld, A. M., Bross, I. D., and Burnett, W. S.,** A prospective study of the relationship between benign breast diseases and breast carcinoma, *Prev. Med.*, 8, 40, 1979.
57. **Gump, F. E., LoGerfo, P., Kister, S., and Habif, D. V.,** Fibrocystic disease of the breast, *N. Engl. J. Med.*, 308, 722, 1983.
58. **Hutchinson, W. B., Thomas, D. B., Hamlin, W. B., Poth, G. J., Peterson, A. V., and Williams, B.,** Risk of breast cancer in women with benign breast disease, *J. Natl. Cancer Inst.*, 65, 13, 1980.
59. **Ketcham, A. S. and Sindelar, W. F.,** Risk factors in breast cancer, *Prog. Clin. Cancer*, 6, 99, 1975.
60. **Horrobin, D. F., Manku, M. S., Brush, M., Callender, K., Preece, P. E., and Mansel, R. E.,** Abnormalities in plasma essential fatty acid levels in women with premenstrual syndrome and with non-malignant breast disease, submitted, 1984.

61. **Karmali, R. A.,** Review: prostaglandins and cancer, *Prostaglandins Med.,* 5, 11, 1980.
62. **Rolland, P. H.,** Involvement of prostaglandins in human glandular epithelium neoplasia with particular reference to the breast and prostate, *J. Chir. (Paris),* 119, 523, 1982.
63. **Rolland, P. H., Martin, P. M., Rolland, A. M., Bourry, M., and Serment, H.,** Benign breast disease: studies of prostaglandin E2, steroids and the thermographic effects of inhibitors of prostaglandin biosynthesis, *Obstet. Gynecol.,* 54, 715, 1979.
64. **Rolland, P. H., Martin, P. M., Jacquemier, J., Rolland, A. M., and Toga, M.,** Prostaglandin production and metabolism in human breast cancer, *Adv. Prostaglandin Thromboxane Res.,* 6, 575, 1980.
65. **Rolland, P. H. and Martin, P. M.,** Prostaglandins and breast cancer, *J. Gynecol. Obstet. Biol. Rep.,* 10, 295, 1981.
66. **Rolland, P. H., Martin, P. M., and Serment, H.,** Steroid hormone receptors and prostaglandin synthesis in fibrocystic breast disease, in *Endocrinology of Cystic Breast Disease,* Angeli, A., Ed., Raven Press, New York, 1983, 203.
67. **Bennett, A.,** Prostaglandins and their synthesis inhibitors in cancer, in *Hormones and Cancer,* Lacobelli, S., Ed., Raven Press, New York, 1980, 515.
68. **Bennett, A., Charlier, E. M., McDonald, A. M., Simpson, J. S., Stamford, I. F., and Zebro, T.,** Prostaglandins and breast cancer, *Lancet,* 2, 624, 1977.
69. **Bennett, A., Charlier, E. M., McDonald, A. M., Simpson, J. S., Stamford, I. F., and Zebro, T.,** Prostaglandins and breast cancer, *Lancet,* 2, 624, 1977.
70. **Feller, N., Malachi, T., and Halbrecht, I.,** Prostaglandin E_2 and cyclic AMP levels in human breast tissues, *J. Cancer Res. Clin. Oncol.,* 93, 275, 1979.
71. **Gunasegaram, R., Loganath, A., Pen, K. L., Chiang, S. C., and Ratnam, S. S.,** Identification of prostaglandins in infiltrating duct carcinoma of the human breast, *IRCS J. Med. Sci.,* 8, 747, 1980.
72. **Kibbey, W. E., Bronn, D. G., and Minton, J. P.,** Prostaglandin synthetase and prostaglandin E_2 levels in human breast carcinoma, *Prostaglandin Med.,* 2, 133, 1979.
73. **Smethurst, M., Bishun, N. P., and Williams, D. C.,** Breast tumour prostaglandin levels and the ability of the tumours to produce prostaglandins in culture, *IRCS Med. Sci.,* 6, 458, 1978.
74. **Stamford, I. F., Carroll, M. A., Civier, A., Hensby, C. N., and Bennett, A.,** Identification of arachidonate metabolites in normal, benign and malignant human mammary tissues, *J. Pharm. Pharmacol.,* 35, 48, 1983.
75. **Fulton, A., Roi, L., and Russo, J.,** Tumor associated prostaglandins in primary breast cancer, *Proc. Am. Assoc. Cancer Res.,* 23, 140, 1982.
76. **Campbell, F. C., Haynes, J., Evans, D. F., Mumford, C., Blamey, R. W., Elston, C. W., and Nicholson, R. I.,** Prostaglandin E_2 synthesis by tumor epithelial cells and oestrogen receptor (ER) status of primary breast cancer, *Langenbecks Archiv. Chirurg.,* 357, 209, 1982.
77. **Rolland, P. H., Martin, P. M., Jacquemier, J., Rolland, A. M., and Toga, M.,** Prostaglandin in human breast cancer: evidence suggesting that an elevated prostaglandin production is a marker of high metastatic potential for neoplastic cells, *J. Natl. Cancer Inst.,* 64, 1061, 1980.
78. **Bennett, A., Houghton, J., Leaper, D. J., and Stamford, I. F.,** Cancer growth, response to treatment and survival time in mice: beneficial effect of the prostaglandin synthesis inhibitor flurbiprofen, *Prostaglandins,* 17, 179, 1979.
79. **Bennett, A., Berstock, D. A., Carroll, M. A., Stamford, I. F., and Wilson, A. J.,** Breast cancer, its recurrence and patient survival in relation to tumor prostaglandins, *Adv. Prostaglandin Thrombox Leukotr. Res.,* 12, 299, 1983.
80. **Brenner, D. E., Harvey, H. A., Lipton, A., and Demers, L.,** A study of prostaglandin E_2, parathormone, and response to indomethacin in patients with hypercalcemia of malignancy, *Cancer,* 49, 556, 1982.
81. **Caro, J. F., Besarab, A., and Flynn, J. T.,** Prostaglandin E and hypercalcemia in breast carcinoma: only a tumor marker?, *Am. J. Med.,* 66, 337, 1979.
82. **Dowsett, M., Easty, G. C., Powles, T. J., Easty, D. M., and Neville, A. M.,** Human breast tumour-induced osteolyses and prostaglandins, *Prostaglandins,* 11, 447, 1976.
83. **Easty, G. C., Dowsett, M., Powles, T. J., Easty, D. M., Gazet, J. C., and Neville, A. M.,** Hypercalcemia in malignant disease, *Proc. R. Soc. Med.,* 70, 191, 1977.
84. **Greaves, M., Ibbotson, K. J., Atkins, D., and Martin, T. J.,** Prostaglandins as mediators of bone resorption in renal and breast tumours, *Clin. Sci.,* 58, 201, 1980.
85. **Minkin, C., Fredericks, R. S., Pokress, S., Rude, R. K., Sharp, C. F., Tong, M., and Singer, F. R.,** Bone receptors and humoral hypercalcemia of malignancy: stimulation of bone resorption in vitro by tumor extracts is inhibited by prostaglandin synthesis inhibitors, *J. Clin. Endocrinol.,* 53, 941, 1981.
86. **Samuel, A. W., Galasko, C. S. B., Rushton, S., and Lacey, E.,** Osteolysis produced by mammary carcinoma: the response to diphosphonates and prostaglandin inhibitors, *Clin. Oncol.,* 8, 86, 1982.
87. **Malachi, T., Chaimoff, C., Feller, N., and Halbrecht, I.,** Prostaglandin E_2 and cyclic AMP in tumor and plasma of breast cancer patients, *J. Cancer Res. Clin. Oncol.,* 102, 71, 1981.

88. **Harvey, P. R. C., Kimura, L. H., Cripps, C., and Hokama, Y.,** Distribution of prostaglandins B, E, and F series in plasma of cancer patients, *J. Med.,* 12, 427, 1981.

89. **Coombes, R. C., Neville, A. M., Bondy, P. K., and Powles, T. J.,** Failure of indomethacin to reduce hydroxyproline excretion or hypercalcemia in patients with breast cancer, *Prostaglandins,* 12, 1027, 1976.

90. **Powles, T. J., Dady, P. J., Williams, J., Easty, C. G., and Coombes, R. C.,** Use of inhibitors of prostaglandin synthesis in patients with breast cancer, *Adv. Prostaglandin Thromboxane Res.,* 6, 511, 1980.

91. **Ritchie, G. A. F.,** The direct inhibition of prostaglandin synthetase of human breast cancer tumour tissue by tamoxifen, *Recent Res. Cancer Res.,* 71, 96, 1980.

92. **Petrakis, N. L., Mason, M. L., Doherty, M., Dupuy, M. E., Sadee, G., and Wilson, C. S.,** Effects of altering diet fat on breast fluid lipids in women, *Fed. Proc.,* 36, 1163, 1977.

93. **Pashby, N. L., Mansel, R. E., Preece, P. E., Hughes, L. E., and Aspinall, J.,** A clinical trial of evening primrose oil in mastalgia, *Br. J. Surg.,* 68, 1, 1981.

94. **Preece, P. E., Hanslip, J. I., Gilbert, L., Walker, D., Pashby, N. L., Mansel, R. E., Evans, B., and Hughes, L. E.,** Evening primrose oil (Efamol) for mastalgia, in *Clinical Uses of Essential Fatty Acids,* Horrobin, D. F., Ed., Eden Press, Montreal, 1982, 147.

95. **Abrams, A. A.,** Use of vitamin E in chronic cystic mastitis, *N. Engl. J. Med.,* 272, 1080, 1965.

96. **London, R. S., Solomon, D. M., London, E. D., Strummer, D., Bankoski, J., and Nair, P. P.,** Mammary dysplasia: clinical response and urinary excretion of 11-deoxy-17-ketosteroids and pregnanediol following α-tocopherol therapy, *Breast,* 4, 19, 1980.

97. **Pearce, M. L. and Dayton, S.,** Incidence of cancer in men on a diet higher in polyunsaturated fat, *Lancet,* 1, 464, 1971.

98. **Ederer, F., Leren, P., Turpeinen, O., and Frandt, I. D.,** Cancer among men on cholesterol-lowering diets. Experience from five clinical trials, *Lancet,* 2, 203, 1971.

99. **Heyden, S.,** Polyunsaturated fatty acids and colon cancer. *Nutr. Met.,* 17, 321, 1974.

100. **WHO/FAO,** Dietary Fats and Oils in Human Nutrition. Report of an Expert Consultation. Food and Agriculture Organization, Rome, 1977.

101. **Stone, K. J., Willis, A., and Hart, M.,** The metabolism of dihomogammalinolenic acid in man, *Lipids,* 14, 174, 1979.

102. **Horrobin, D. F., Huang, Y-S., Cunnane, S. C., and Manku, M. S.,** Essential fatty acids in plasma, red blood cells and liver phospholipids in common laboratory animals as compared to humans, *Lipids,* in press, 1984.

103. **Horrobin, D. F. and Manku, M. S.,** How do polyunsaturated fatty acids lower plasma cholesterol levels?, *Lipids,* 18, 558, 1983.

104. **Knazek, R. A., Liu, S. C., Bodwin, J. S., and Vonderhaar, B. K.,** Requirement of essential fatty acids in the diet for the development of the mouse mammary gland, *J. Natl. Cancer Inst.,* 64, 377, 1980.

105. **Knazek, R. A. and Liu, S. C.,** Effects of dietary essential fatty acids on murine mammary gland development, *Cancer Res.,* 41, 3750, 1981.

106. **Knazek, R. A., Liu, S. C., St. Amand, L. M., Dave, J. R., and Christy, R. J.,** Cholesterol accentuates the effect of unsaturated fatty acid deficiency on mammary gland development, *Proc. Am. Assoc. Cancer Res.,* 22, 10, 1981.

107. **Abraham, S. and Hillyard, L. A.,** Effect of dietary 18-carbon fatty acids on growth of transplantable mammary adenocarcinomas in mice, *J. Natl. Cancer Inst.,* 71, 601, 1983.

108. **Mead, J. F., and Fulco, A. J.,** *The Unsaturated and Polyunsaturated Fatty Acids in Health and Disease,* Charles C Thomas, Springfield, Ill., 1976.

109. **Holman, R. T.,** Essential fatty acid deficiency, in *Progress in the Chemistry of Fats and Other Lipids,* Holman, R. T., Ed., Pergamon Press, Elmsford, N.Y., 1966, 275.

110. **Silverstone, H. and Tannenbaum, A.,** The effect of proportion of dietary fat on the rate of formation of mammary carcinoma in mice, *Cancer Res.,* 10, 448, 1950.

111. **Benson, T., Lev, M., and Grand, C. G.,** Enhancement of mammary fibroadenomas in the female rat by a high fat diet, *Cancer Res.,* 16, 135, 1956.

112. **Carroll, K. K. and Hopkins, G. J.,** Dietary polyunsaturated fat versus saturated fat in relation to mammary carcinogenesis, *Lipids,* 14, 155, 1979.

113. **Carroll, K. K., Hopkins, G. J., and Kennedy, T. G.,** Essential fatty acids in relation to mammary carcinogenesis, *Prog. Lipid Res.,* 20, 685, 1981.

114. **Davidson, M. B. and Carroll, K. K.,** Inhibitory effect of a fat-free diet on mammary carcinogenesis in rats, *Nutr. Cancer,* 3, 207, 1982.

115. **Gammal, E. B., Carroll, K. K., and Plunkett, E. R.,** Effects of dietary fat on mammary carcinogenesis by 7,12-dimethylbenz(α)anthracene in rats, *Cancer Res.,* 27, 1737, 1967.

116. **Gammal, E. B., Carroll, K. K., and Plunkett, E. R.,** Effects of dietary fat on the uptake and clearance of 7,12-dimethylbenz(α)anthracene by rat mammary tissue. *Cancer Res.,* 28, 384, 1968.

117. **Hopkins, G. J. and Carroll, K. K.,** Relationship between amount and type of dietary fat in promotion of mammary carcinogenesis induced by 7,12-dimethylbenz(α)anthracene, *J. Natl. Cancer Inst.,* 62, 1009, 1979.

118. **Hopkins, G. J., Kennedy, T. G., and Carroll, K. K.,** Polyunsaturated fatty acids as promoters of mammary carcinogenesis induced in Sprague-Dawley rats by 7,12-dimethylbenz(α)anthracene, *J. Natl. Cancer Inst.,* 66, 517, 1981.

119. **St. Angelo, A. J. and Ory, R. L.,** Lipoxygenase inhibition by a naturally occurring monoenoic acid, *Lipids,* 19, 34, 1984.

120. **Ip, C.,** Ability of dietary fat to overcome the resistance of mature female rats to 7,12-dimethylbenz(α)anthracene-induced mammary tumorigenesis, *Cancer Res.,* 40, 2785, 1980.

121. **Ip, C. and Sinha, D.,** Neoplastic growth of carcinogen-treated mammary transplants as influenced by fat intake of donor and host, *Cancer Lett.,* 11, 277, 1981.

122. **Chan, P-C., Ferguson, K. A., and Dao, T. L.,** Effects of different dietary fats on mammary carcinogenesis, *Cancer Res.,* 43, 1079, 1983.

123. **Dao, T. L. and Chan, P-C.,** Hormones and dietary fat as promoters in mammary carcinogenesis, *Environ. Health Perspect.,* 50, 219, 1983.

124. **Dayton, S., Hashimoto, S., and Wollman, J.,** Effect of high-oleic and high-linoleic safflower oils on mammary tumors induced in rats by 7,12-dimethylbenz(α)anthracene, *J. Nutr.,* 107, 1353, 1977.

125. **Rees, E. D. and Ackermann, H.,** Mammary gland fatty acids in rats of different susceptibility to mammary carcinoma induction, *Proc. Soc. Exp. Biol. Med.,* 127, 106, 1968.

126. **To, D., Manning, L., and Carpenter, M.,** Prostaglandin precursor fatty acids of rat mammary gland: effect of diet, physiological state and neoplasia, *Fed. Proc.,* 37, 358, 1978.

127. **Leung, B. S. and Sun, G. Y.,** Acyl group composition of membrane phospholipids in mammary tissues and carcinoma induced by dimethylbenz(α)anthracene, *Proc. Soc. Exp. Biol. Med.,* 152, 671, 1976.

128. **Hillyard, L., Rao, G. A., and Abraham, S.,** Effect of dietary fat on fatty acid composition of mouse and rat mammary adenocarcinomas, *Proc. Soc. Exp. Biol. Med.,* 163, 376, 1980.

129. **Hillyard, L. A. and Abraham, S.,** Effect of dietary polyunsaturated fatty acids on growth of mammary adenocarcinomas in mice and rats, *Cancer Res.,* 39, 4430, 1979.

130. **Hopkins, G. J. and West, C. E.,** Possible roles of dietary fats in carcinogenesis, *Life Sci.,* 19, 1103, 1976.

131. **Hopkins, G. J. and West, C. E.,** Effect of dietary polyunsaturated fat on the growth of a transplantable adenocarcinoma in C3H-A fB mice, *J. Natl. Cancer Inst.,* 58, 753, 1977.

132. **Hopkins G. J., Hard, G. C. and West, C. E.,** Carcinogenesis induced by 7,12-dimethylbenz(α)anthracene in C3H-A fB mice: influence of different dietary fats, *J. Natl. Cancer Inst.,* 60, 849, 1978.

133. **Brown, R. R.,** Effects of dietary fat on incidence of spontaneous and induced cancer in mice, *Cancer Res.,* 41, 3741, 1981.

134. **Gridley, D. S., Kettering, J. D., Slater, J. M., and Nutter, R. L.,** Modification of spontaneous mammary tumors in mice fed different sources of protein, fat and carbohydrate, *Cancer Lett.,* 19, 133, 1983.

135. **Tinsley, I. J., Schmitz, J. A., and Pierce, D. A.,** Influence of dietary fatty acids on the incidence of mammary tumors in the C3H mouse, *Cancer Res.,* 41, 1460, 1981.

136. **Ghayur, T. and Horrobin, D. F.,** Effects of essential fatty acids in the form of evening primrose oil on the growth of the rat R3230AC transplantable mammary tumour, *IRCS J. Med. Sci.,* 9, 582, 1981.

137. **Horrobin, D. F.,** The reversibility of cancer: the relevance of cyclic AMP, calcium, essential fatty acids and prostaglandin E_1, *Med. Hypoth.,* 6, 469, 1980.

138. **Dippenaar, N., Booyens, J., Fabri, D., and Katzeff, I. E.,** The reversibility of cancer: evidence that malignancy in melanoma cells is gamma-linolenic acid deficiency dependent, *S. Afr. Med. J.,* 62, 505, 1982.

139. **Leary, W. P., Robinson, K. M., Booyens, J., and Dippenaar, N.,** Some effects of gamma-linolenic acid on cultured human esophageal carcinoma cells, *S. Afr. Med. J.,* 62, 681, 1982.

140. **Booyens, J., Engelbrecht, P., LeRoux, S., Louwrens, C. C., van Der Merwe, C. F., and Katzeff, I. E.,** Some effects of the essential fatty acids, linoleic and alpha-linolenic acid, and of their metabolites gamma-linolenic acid, arachidonic acid, eicosapentaenoic acid, and of prostaglandins A_1 and E_1 on the proliferation of human osteogenic sarcoma cells in culture, *Prostaglandins Leukotrienes Med.,* in press.

141. **Kidwell, W. R., Monaco, M. E., Wicha, M. S., and Smith, G. S.,** Unsaturated fatty acid requirements for growth and survival of a rat mammary tumor cell line, *Cancer Res.,* 38, 4091, 1978.

142. **Wicha, M. S., Liotta, L. A., and Kidwell, W. R.,** Effects of free fatty acids on the growth of normal and neoplasia rat mammary epithelial cells, *Cancer Res.,* 39, 426, 1979.

143. **Barnes, D. W. and Sato, G.,** Growth of a human breast cancer cell line (MCF-7) in a defined serum-free medium, *In Vitro,* 15, 195, 1979.

144. **Rudland, P. S., Davies, A. T. and Warburton, M. J.,** Prostaglandin-induced differentiation or dimethylsulfoxide-induced differentiation: reduction of the neoplastic potential of a rat mammary tumor stem cell line, *J. Natl. Cancer Inst.,* 69, 1083, 1982.

145. **Svec, J., Svec, P., Halcak, L., and Thurzo, V.,** Role of natural prostaglandins in the control of murine mammary tumor virus expression, *J. Cancer Res. Clin. Oncol.,* 103, 55. 1982.

146. **Huang, Y-S., Manku, M. S., and Horrobin, D. F.,** The effects of dietary cholesterol on blood and liver polyunsaturated fatty acids and on plasma cholesterol in rats fed various types of fatty acid diet, *Lipids,* in press, 1984.

147. **Miyamoto-Tlaven, M. J., Hillyard, L. A., and Abraham, S.,** Influence of dietary fat on the growth of mammary ducts in BALB/c mice, *J. Natl. Cancer Inst.,* 67, 179, 1981.

148. **Rao, G. A. and Abraham, S.,** Dietary alteration of fatty acid composition of lipid classes in mouse mammary adenocarcinoma, *Lipids,* 10, 641, 1975.

149. **Rao, G. A. and Abraham, S.,** Enhanced growth rate of transplanted mammary adenocarcinoma induced in C3H mice by dietary linoleate, *J. Natl. Cancer Inst.,* 56, 431, 1976.

150. **Hillyard, L. A. and Abraham, S.,** Effect of dietary polyunsaturated fatty acids on growth of mammary adenocarcinomas in mice and rats, *Cancer Res.,* 39, 4430, 1979.

151. **Abraham, S. and Hillyard, L. A.,** Effect of dietary 18-carbon fatty acids on growth of transplantable mammary adenocarcinomas in mice, *J. Natl. Cancer Inst.,* 71, 601, 1983.

152. **Cave, W. T. and Erickson-Lucas, M. J.,** Effect of dietary lipids on lactogenic hormone receptor binding in rat mammary tumors, *J. Natl. Cancer Inst.,* 68, 319, 1982.

153. **Terada, N., Leiderman, L. J., and Oka, T.,** The interaction of cortisol and prostaglandins on the phenotypic expression of the α-lactalbumin gene in the mouse mammary gland in culture, *Biochem. Biophys. Res. Commun.,* 111, 1059, 1983.

154. **Rillema, J. A.,** Activation of casein synthesis by prostaglandins plus spermidine in mammary gland explants of mice, *Biochem. Biophys. Res. Commun.,* 70, 45, 1976.

155. **Rillema, J. A.,** Effects of prostaglandins on RNA and casein synthesis in mammary gland explants of mice, *Endocrinology,* 99, 490, 1976.

156. **Horrobin, D. F.,** Cellular basis of prolactin action, *Med. Hypoth.,* 5, 599, 1979.

157. **Carpenter, M. P., Robinson, R. D., and Thuy, L. P.,** Prostaglandin synthesis and prostaglandin E-9-ketoreductases in normal and neoplastic rat mammary gland, *Fed. Proc.,* 36, 767, 1977.

158. **Fulton, A. M.,** Prostaglandin levels in mammary tumors of varying malignant potential, *Proc. Am. Assoc. Cancer Res.,* 22, 61, 1981.

159. **Tan, W. C., Privett, O. S. and Goldyne, M. E.,** Studies of prostaglandins in rat mammary tumors induced by 7,12-dimethylbenz(α)anthracene, *Cancer Res.,* 34, 3229, 1974.

160. **Busse, E., Helmholz, M., and Magdon, E.,** Altersbedingte veranderungen des cAMP-und prostaglandingehaltes im Gehirn und in den Lymphozyten von spontan mammakarzinombildenden C3H-Mausen, *Onkologie,* 4, 328, 1981.

161. **Foecking, M. K., Kibbey, W. E., Abou-Issa, H., Matthews, R. H., and Minton, J. P.,** Hormone dependence of 7,12-dimethylbenz(α)anthracene-induced mammary tumor growth: correction with prostaglandin E_2 content, *J. Natl. Cancer Inst.,* 69, 443, 1982.

162. **Kibbey, W. E., Bronn, D. G., and Minton, J. P.,** Prostaglandins and metastasis, *Lancet,* 1, 101, 1978.

163. **Abou-Issa, H., Minton, J. P., Foeching, M. K., and Bronn, D. G.,** Prostaglandin E binding in hormonally progressive rat mammary tumors, *Fed. Proc.,* 42, 518, 1983.

164. **Karmali, R. A., Thaler, H. T., and Cohen, L. A.,** Prostaglandin concentrations and prostaglandin synthetase activity in N-nitrosomethylurea-induced rat mammary adenocarcinoma, *Rev. J. Cancer Clin. Oncol.,* 19, 817, 1983.

165. **Puck, T. T.,** Cyclic AMP, the microtubule microfilament system and cancer, *Proc. Natl. Acad. Sci. U.S.A.,* 74, 4491, 1977.

166. **Johnson, G. S., Pastan, I., Peery, C. V., Otten, J., and Willingham, M.,** The role of prostaglandins in the regulation of growth and morphology of transformed fibroblasts, in *Prostaglandins in Cellular Biology,* Ramwell, P. W., Pharriss, and B. B. Rloreen, Eds., Plenum, N.Y., 1972, 195.

167. **Rao, G. A. and Abraham, S.,** Reduced growth rate of transplantable mammary adenocarcinoma in C3H mice fed eicosa-5,8,11,14-tetraynoic acid, *J. Natl. Cancer Inst.,* 58, 445, 1977.

168. **Fulton, A. M. and Heppner, G.,** Murine mammary tumor growth inhibition with indomethacin, *Proc. Am. Assoc. Cancer Res.,* 23, 21, 1983.

169. **McCormick, D. L., Faikus, S. A., and Moon, R. C.,** Modification of rat mammary carcinogenesis by flurbiprofen, an inhibitor of prostaglandin biosynthesis, *Fed. Proc.,* 42, 783, 1983.

170. **Carter, C. A., Milholland, R. J., Shea, W., and Ip, M. M.,** Effect of the prostaglandin synthetase inhibitor indomethacin on 7,12-dimethylbenz(a)-anthracene-induced mammary tumorigenesis in rats fed different levels of fat, *Cancer Res.,* 43, 3559, 1983.

171. **Ip, M. M., Carter, C.A., Milholland, R. J., and Shea, W. K.,** Effect of indomethacin on 7,12-dimethylbenz(a)anthracene induced mammary tumorigenesis in rats fed different levels of fat, *Proc. Am. Assoc. Cancer Res.,* 24, 97, 1983.

172. **Jacobson, H. I.,** Oncolytic action of prostaglandins, *Cancer Chemother, Rep.,* 58, 503, 1974.

173. **Jubiz, W., Frailey, J., and Smith, J. B.,** Inhibitory effect of prostaglandin $F_{2\alpha}$ on the growth of a hormone-dependent rat mammary tumour, *Cancer Res.,* 39, 998, 1979.

174. **Bennett, A., Berstock, D. A., and Carroll, M. A.,** Increased survival of cancer-bearing mice treated with inhibitors of prostaglandin synthesis alone or with chemotherapy, *Br. J. Cancer,* 45, 762, 1982.

175. **Leaper, D. J., French, B. T., and Bennett, A.,** Breast cancer and prostaglandins: a new approach to treatment, *Br. J. Surg.,* 66, 683, 1979.

176. **Homa, S. T., Conroy, D. M., and Smith, A. D.,** Unsaturated fatty acids stimulate the formation of lipoxygenase and cyclo-oxygenase products in rat spleen lymphocytes, *Prostaglandins, Leukotrienes,* Med, in press, 1984.

177. **Horrobin, D. F.,** *Prolactin 7,* Eden Press, Montreal, 1979.

178. **Chan, P-C. and Cohen, L. A.,** Effect of dietary fat, antiestrogen, and antiprolactin on the development of mammary tumors in rats, *J. Natl. Cancer Inst.,* 52, 25, 1974.

179. **Chan, P-C. and Cohen, L. A.,** Dietary fat and growth promotion of rat mammary tumors, *Cancer Res.,* 35, 3384, 1975.

180. **Chan, P-C., Didato, F., and Cohen, L. A.,** High dietary fat, elevation of rat serum prolactin and mammary cancer, *Proc. Soc. Exp. Biol. Med.,* 149, 133, 1975.

181. **Chan, P-C., Head, J. F., Cohen, L. A., and Wynder, E. L.,** Influence of dietary fat on the evolution of mammary tumors by N-nitrosomethylurea: associated hormone changes and differences between Sprague-Dawley and F344 rats, *J. Natl. Cancer Inst.,* 59, 1279, 1977.

182. **Chan, P-C., Head, J. F., Cohen, L. A., and Wynder, E. L.,** Effect of high fat diet on serum prolactin levels and mammary cancer development in ovariectomized rats, *Proc. Am. Assoc. Cancer Res.,* 18, 189, 1977.

183. **Cohen, L. A., Chan, P. C., and Wynder, E. L.,** The role of a high fat diet in enhancing the development of mammary tumors in ovariectomized rats, *Cancer,* 47, 66, 1981.

184. **Ip, C., Yip, P., and Bernardis, L. L.,** Role of prolactin in the formation of dimethylbenz(a)anthracene-induced mammary tumors by dietary fat, *Cancer Res.,* 40, 374, 1980.

185. **Aylsworth, C. F., Van Vugt, D. A., Sylvester, P. W., and Meites, J.,** A direct mechanism by which high fat diets stimulate mammary tumor development, *Proc. AACR ASCO,* 22, 12, 1981.

186. **King, M. M. and Otto, P.,** Null effect of BHA and α-tocopherol on 7,12-dimethylbenz(α)anthracene-induced mammary tumors in rats fed different levels and types of dietary fat, *Proc. Am. Assoc. Cancer Res.,* 20, 227, 1979.

187. **King, M. M. and McCay, P. B.,** DMBA-induced mammary tumor incidence: effect of propyl gallate supplementation in purified diets containing different types and amounts of fat, *Proc. Am. Assoc. Cancer Res.,* 21, 113, 1980.

188. **King, M. M., Bailey, D, M., Gibson, D. D., Pitha, J. V., and McCay, P. B.,** Effect of anti-oxidant in diets containing different types and amounts of fat on mammary tumor incidence induced by a single dose of 7,12-dimethylbenzanthracene, *Fed. Proc.,* 36, 1148, 1977.

189. **King, M. M., Bailey, D. M., Gibson, D. D., Pitha, J. V., and McCay, P. B.,** Incidence and growth of mammary tumors induced by dimethylbenzanthracene as related to the dietary content of fat and anti-oxidant, *J. Natl. Cancer Inst.,* 63, 657, 1979.

190. **McCay, P. B., King, M., Rikans, L. E., and Pitha, J. V.,** Interactions between dietary fats and antioxidants on DMBA-induced mammary carcinomas and on AAF-induced hyperplastic nodules and tumors, *J. Environ. Pathol. Toxicol.,* 3, 451, 1980.

191. **St. Angelo, A. J. and Ory, R. L.,** Lipoxygenase inhibition by a naturally occurring monoenoic acid, *Lipids,* 19, 34, 1984.

192. **Ip, C.,** Factors influencing the anticarcinogenic efficacy of selenium in dimethylbenz(a)anthracene-induced mammary tumorigenesis in rats, *Cancer Res.,* 41, 2683, 1981.

193. **Ip, C. and Sinha, D. K.,** Enhancement of mammary tumorigenesis by dietary selenium deficiency in rats with a high polyunsaturated fat intake, *Cancer Res.,* 41, 31, 1981.

194. **Ip. C. and Sinha, D.,** Anticarcinogenic effect of selenium in rats treated with dimethylbenz(a)anthracene and fed different levels and types of fat, *Carcinogenesis,* 2, 435, 1981.

195. **Field, E. J. and Shenton, B. K.,** Inhibitory effect of unsaturated fatty acids on lymphocyte-antigen interaction with special reference to multiple sclerosis, *Acta Neurol. Scand.,* 52, 121, 1975.

196. **Mertin, J. and Hunt, R.,** Influence of polyunsaturated fatty acids on survival of skin allographs and tumor incidence in mice, *Proc. Natl. Acad. Sci. U.S.A.,* 73, 928, 1976.

197. **Mertin, J. and Stackpoole, A.,** Anti-PGE antibodies inhibit in vivo development of cell mediated immunity, *Nature (London),* 294, 456, 1981.

198. **Newberne, P. M.,** Dietary fat, immunological response and cancer in rats, *Cancer Res.,* 41, 3783, 1981.

199. **Kunkel, S. L., Ward, P. A., and Zurier, R. B.,** Modulation of inflammatory responses by prostaglandins and essential fatty acids, *Fed. Proc.,* 40, 344, 1981.

200. **Plescia, O. J., Smith, A. H., and Grenwich, K.,** Subversion of immune system by tumor cells and role of prostaglandins, *Proc. Natl. Acad. Sci. U.S.A.,* 72, 1848, 1975.

201. **Goodwin, J. S., Husby, G., and Williams, R. C.,** Prostaglandin E and cancer growth, *Cancer Immunol. Immurother.*, 8, 3, 1980.
202. **Poleshuck, L. C. and Strausser, H. R.,** Immune-complex induced prostaglandin production by monocytes of normal human subjects and cancer patients, *Prostaglandins Med.*, 4, 363, 1980.
203. **Cameron, D. J. and O'Brien, P.,** Relationship of the suppression of macrophage mediated tumor cyto-toxicity in conjunction with secretion of prostaglandin from the macrophages of breast cancer patients, *Int. J. Immunopharmacol.*, 4, 445, 1982.
204. **Wagner, D. A., Naylor, P. H., Kim, U., Shea, W., Ip, C., and Ip, M. M.,** Interaction of dietary fat and the thymus in the induction of mammary tumors by 7,12-dimethylbenz(*a*)anthracene, *Cancer Res.*, 42, 1266, 1982.
205. **Clinton, S. K., Mulloy, A. L., Truex, C. R., and Visek, W. J.,** Mammary tumors and carcinogen metabolizing enzymes in rats fed corn oil and beef tallow, *J. Nutr.*, 109, 18, 1979.
206. **Fox, O. F., Kishore, G. S., and Carubelli, R.,** Sialic acid metabolism in rats undergoing chemically-induced mammary gland carcinogenesis in specific dietary states, *Cancer Lett.*, 7, 251, 1979.
207. **Kollmorgen, G. M., King, M. M., Kosanke, S. D., and Do, C.,** Influence of dietary fat and indomethacin on the growth of transplantable mammary tumors in rats, *Cancer Res.*, 43, 4714, 1983.
208. **Goodman, D. G., Ward, J. M., Squire, R. A., Paxton, M. B., Reichardt, W. D., Chu, K. C., and Linhart, M. S.,** Neoplastic and non-neoplastic lesions in aging Osborne-Mendel rats, *Toxicol. Appl. Pharmacol.*, 55, 433, 1980.

Chapter 8

DIET AND CANCER OF THE PROSTATE: EPIDEMIOLOGIC AND EXPERIMENTAL EVIDENCE

Maarten C. Bosland

TABLE OF CONTENTS

I. INTRODUCTION

Cancer of the prostate is one of the main human cancers in the Western world. It is the second leading cause of death due to cancer in men in countries such as the U.S.[1] Prostatic cancer is only important at old age, i.e., in men over 65 to 70 years of age.[2,3] World-wide, age-specific prostatic cancer death and incidence rates have, with few exceptions, increased over time (Table 1).[4-7] Notwithstanding the relative importance of the disease, both in terms of mortality and morbidity and as a clinical entity, the etiology is essentially unknown. Recent research efforts have resulted in improved diagnostic and therapeutic procedures and a better understanding of the biology of the tumor, its behavior, and particularly the endocrinology of prostate cancer.[8] Little insight in the etiology is derived from experimental studies. The epidemiology of the disease, on the other hand, has indicated that dietary, sexual, and genetic and racial factors are of possible etiological importance.[2,3,9-11]

In this chapter a review will be presented of the evidence available to date in favor of a dietary etiology. Epidemiology will receive major attention. Nondietary factors of possible importance in the etiology of cancer of the prostate will only be covered in order to estimate their contribution to the cancer burden relative to that of dietary factors. For more information on nondietary factors in prostatic carcinogenesis, we refer to some recent reviews.[2,3,9-11]

II. DESCRIPTIVE EPIDEMIOLOGY

A. Geographic Variation and Migrant Studies

The occurence of cancer of the prostate, in terms of both mortality and incidence, shows a considerable geographic variation (Table 1).[4-7] Rates are generally high in the U.S., Canada, and northwest European countries. Lower rates are found in east and south Europe, and in South American countries. Extremely low rates are found in Japan and some Central American and southeast Asian countries. Generally, rates seem to be higher in countries with a high degree of development (affluence), and low in less developed countries. However, there are important exceptions, such as Japan where rates are extremely low (Table 1).

Large-scale migration from low-risk areas, in particular from Asiatic and east European countries to the U.S., has given the unique opportunity to study mortality and morbidity patterns in these migrant populations in comparison to the population of their homeland and their host country. Thus, it is possible to discriminate between environmental and genetic factors in the etiology of the disease. In all populations that have migrated from a low-risk area to the U.S., a significantly higher death and incidence rate is found than in their homelands, but the rates are still slightly to markedly lower than those of U.S. native whites.[3] This has been found for east European Caucasians as well as for Japanese and Chinese migrants (Table 1).[3]

Thus, populations migrating from a low-risk area to a high-risk area tend to acquire the high risk for prostate cancer of their new environment. Therefore, endogenous factors, i.e., genetically or constitutionally determined differences in susceptability, are clearly less important than exogenous (environmental) determinants.

B. Racial Differences and Familial Aggregation

Death rates in black men in the U.S. are considerably higher than white rates.[2,3] Differences in survival or accessibility of medical care cannot explain this impressive black/white difference.[2,3]

Interestingly, in black men living in Africa incidence and mortality are very low in comparison with U.S. blacks.[2,12] The differences in risk between the black and white male population in the U.S. are perhaps partly due to genetically controlled differences in susceptibility for the disease. On the other hand, the rates among U.S. blacks may be influenced by environmental factors as well.

Table 1
AGE-ADJUSTED DEATH AND INCIDENCE RATES FOR CANCER
OF THE PROSTATE FOR DIFFERENT COUNTRIES AND AREAS[a]

	Death rates		Incidence rates[b]		
	1973	1976	1968—1972	1973—1977	
Switzerland	19.84	18.73	29.9	36.3	(Geneva)
Sweden	19.53	21.03	38.8	44.4	
Norway	15.41	19.33	33.1	38.9	
Urban			36.3	42.4	
Rural			31.0	36.4	
F.R. Germany	14.74	15.28	22.9	28.5	(Hamburg)
Netherlands	14.50	16.15			
France	14.41	14.83		23.0	(Bas-rhin)
Urban				25.7	(Bas-rhin)
Rural				20.9	(Bas-rhin)
Hungary	14.23	15.43	9.1	10.1	(Szabolcs)
Urban				8.9	(Szabolcs)
Rural				10.4	(Szabolcs)
U.S.	14.18	14.50			
White			44.6	47.4	(Bay Area)
Black			77.0	92.2	(Bay Area)
Chinese			18.2	18.6	(Bay Area)
Japanese			12.7		(Bay Area)
Spanish			34.3	38.7	(New Mexico)
Other white			50.1	54.6	(New Mexico)
American Indian			27.5	31.6	(New Mexico)
White			36.1	41.4	(Detroit)
Black			67.1	73.2	(Detroit)
Hawaiian			19.8	42.5	(Hawaii)
White			42.3	59.7	(Hawaii)
Chinese			17.8	25.8	(Hawaii)
Filipino			14.0	30.5	(Hawaii)
Japanese			24.6	35.9	(Hawaii)
Cuba	14.13		18.0	19.9	
Finland	13.94		22.7	27.2	
Urban				32.1	
Rural				23.2	
Canada	13.65	13.43	32.4	38.1	(Alberta)
Trinidad and Tobago	13.42				
Denmark	13.15	13.66	32.0	23.6	
England and Wales	11.53	11.98		19.2	(Northwestern)
Urban				20.8	(Northwestern)
Rural				12.8	(Northwestern)
Chile	11.43	11.45			
Spain	11.38	12.48	17.7	20.7	(Zaragoza)
Urban				22.3	(Zaragoza)
Rural				19.2	(Zaragoza)
Puerto Rico	10.85		21.4	25.0	
Venezuela	9.76	11.24			
Paraguay	9.20	8.81			
Poland	8.00	8.31			
Urban			14.6	15.6	(Warsaw, city)
Rural			9.4	11.7	(Warsaw, rural)
Romania	7.92	8.12			
Costa Rica	7.65	9.58			
Yugoslavia	7.10	9.20	16.8	15.8	(Slovenia)

Table 1 (continued)
AGE-ADJUSTED DEATH AND INCIDENCE RATES FOR CANCER
OF THE PROSTATE FOR DIFFERENT COUNTRIES AND AREAS[a]

	Death rates		Incidence rates[b]		
	1973	1976	1968—1972	1973—1977	
Greece	6.90	6.17			
Bulgaria	5.99	5.89			
Dominican Republic	5.73				
Mexico	5.07				
Ecuador	4.82				
Japan	2.18	2.28	2.7	3.4	(Osaka)
			2.7	4.9	(Miyagi)

[a]   Rates per 100,00; adjusted to "world population"; sources: Segi[4,5] and Waterhouse et al.[6,7]
[b]   The area for which the incidence rates are presented are given in brackets.

Cancer of the prostate appears to occur more frequently in fathers and brothers of prostatic cancer patients than in unrelated men as summarized by Mandel and Schuman.[3] This familial aggregation is a consistent finding, but the number of reports is small and the excess of risk found is not very high.[3] Familial aggregation may be an expression of a genetic component, but may also be linked to a common environmental exposure of close relatives.[10]

Thus, genetic factors may influence prostatic cancer risk to some extent in certain individuals or ethnic entities. It is not likely, however, that genetic mechanisms account for a substantial proportion of the cases in the Western world.[10]

C. Geographic Pathology of Latent Carcinoma of the Prostate

In a remarkably high frequency, carcinoma of the prostate is found in material from routine autopsies and in surgical specimens from patients with benign hyperplasia of the prostate, when careful histological examination is performed.[13,14] The incidence of this latent prostatic cancer can be as high as 50% in aged men.[13,14] Latent prostate cancer cannot be detected clinically and is asymptomatic.

In general, the prevalence of latent carcinoma of the prostate appears to be remarkably similar in populations that differ in risk for prostatic cancer,[12,13-16] There is only a slight tendency that the prevalence of latent carcinoma is somewhat higher in populations at high risk for clinical prostatic cancer than in low-risk populations. This is due to the fact that the prevalence of large, infiltratively growing (LIT) latent tumors is clearly higher in these high risk populations. Smaller, noninfiltrating (LNT) tumors occur equally or even slightly less frequently in high-risk populations as compared to low-risk groups.

It is attractive to hypothesize that the environmental factors responsible for the geographic differences in prostatic cancer risk act primarily at the phase of progression from small, noninfiltrating tumor to larger, infiltratively growing cancer.[15] This view presupposes that LNT lesions are precursors of LIT tumors. Whether this is true, is not known and difficult to investigate. Some support for this assumption can be derived from the finding in some studies[15,16] that, while the prevalence of LIT tumors increases with increasing risk for clinical prostate cancer, the prevalence of LNT lesions seem to decrease. On the other hand, LNT and LIT tumors could represent different biological entities; or only a certain proportion of the LNT lesions, morphologically indistinguishable from other LNT lesions, could be precursors for LIT lesions. The finding in a study by Guileyardo et al.[16] of an apparent subset of LIT lesions in U.S. blacks that are appreciably bigger than those found in white men, could point to a certain proportion of the LIT lesions in high-risk populations having a high potential for agressive and fast growth. Finally, clinical cancer may arise from LIT lesions, but might also develop as a separate process. Rather strong support for the former view comes from the comparable geographic variation in the occurrence of clinical cancer and large LIT tumors.

D. Environmental Factors

There are no indications that the chemical environment is very important for prostatic cancer risk[3,11] in the general population. Certain occupational exposures, such as to cadmium[11] are perhaps related to prostatic cancer.[3]

Lifestyle factors on the other hand are more clearly associated with prostatic cancer risk, as will be shown below. Lifestyle is a composite of cultural, i.e., group behavioral patterns, and individual characteristics. The more important components of lifestyle are smoking, dietary habits, and sexual behavior. There is no evidence that smoking influences risk for cancer of the prostate.[17-21] Sexual aspects will be briefly summarized below. Diet will receive in-depth attention in subsequent sections.

In general, married men have a slightly higher risk for prostate cancer than single men, and widowed and divorced men a somewhat higher risk than married men.[3] There are also studies that did not show a relation between marital status and prostate cancer risk.[3,17,21] Fertility may be positively associated with prostatic cancer risk.[3,21] These associations are possibly dependent on racial and/or cultural background.

The possible relation with various sexual factors has been investigated in a number of case-control studies.[3,21-27] The general impression from these studies is that, as compared to controls, cases (1) may have an earlier onset of sexual activity in any sense, (2) may show a higher sexual drive, especially at young age, (3) seem to have, notwithstanding their high sexual drive, a lower frequency of intercourse, especially at later age, or, as shown in some studies, they have a higher frequency till an age of about 45 to 50 and a lower frequency thereafter. There is no perfect consistency in the data available with regard to these aspects, and the relative risks observed are rather low.[3] Consistent is the observation that cases slightly more often report a history of venereal disease.[3,23,24]

Thus, it appears that factors associated with sexual activity or sexual behavior may be involved in prostatic carcinogenesis in men. The associations found are rather weak. Hormonal factors, as affected by or affecting sexual activity, or constitutional factors have been postulated to be responsible.[3,10] Also sexually transmissible factors have been implicated,[3,10] which supported by the finding that venereal disease has occurred more frequently in cases than in controls. Certain viruses have been held responsible, but any hard evidence is lacking.[3,8,10] On the other hand, the secretory activity of the gland, prostatic epithelial cell turn-over, and the interaction of the prostatic grandular cells with factors in the luminal fluid and the flow of this fluid[28] could be important in prostatic cancer risk.

III. DIETARY FACTORS

The international geographic patterns in the occurrence of prostatic cancer parallel those of breast cancer in females and colon cancer in both sexes.[6,7,17,29] Also, migrant studies reveal similar trends for these three sites.[17,30] These similarities in epidemiological patterns between prostate cancer and cancer of the colon and the female breast, both strongly related to dietary factors in this book suggest a dietary association.

There are, however, important exceptions to this general impression: In Hawaii, Caucasians have twice the prostate cancer incidence of the other ethnic groups, while Hawaiian natives have breast cancer rates equal to those of Caucasians, but their colon cancer rates are lower.[31] Chinese and Japanese in Hawaii have lower breast and prostate cancer rates than Caucasians, but equal colon cancer rates.[31] In the U.S. black population, colon and breast cancer occur as frequently, or maybe even slightly less frequently than in U.S. whites.[1] Breast and in particular colon cancer rates are clearly lower in Californian Seventh-Day Adventists (SDA) and Utah Mormons (the Church of Jesus Christ of Latter Day Saints, LDS) than in the general U.S. white population, while this is not true for prostate cancer.[32-34] These exceptions may indicate that dietary factors are not as important in prostatic carcinogenesis as they are for breast and colon cancer, and that there are also other major determinants of prostatic cancer risk.

A. Epidemiology

1. Correlation Studies

a. International (Between-Country) Correlation Studies

International data of the estimated per capita consumption of various food commodities, including alcoholic and nonalcoholic beverages, and the derived per capita consumption of certain macronutrients and energy have been compared with international mortality figures for cancer of the prostate.[18,35-40] In these different studies, the sources of mortality and food consumption data used are remarkably similar.

In Table 2, the correlation data from these studies are summarized, and for each item we arbitrarily evaluated strength and consistency of the data.

Schrauzer and co-workers[41,42] calculated per capita selenium intake from known data on the average content of different foods and food consumption data from the FAO over 1964 to 1966. They did not take into account regional differences in selenium content, but they did exclude New Zealand which is a known selenium-deficient area from their analysis. They calculated a significant correlation coefficient of -0.65 for apparent per capita selenium intake and prostate cancer mortality over the period 1964 to 1966[5] for 27[42] or 28[41] countries. They also measured the selenium content in pooled whole blood obtained from bloodbanks in 22 different countries and found a significant inverse correlation with prostate cancer mortality.[42]

b. Intranational (Within-Country) Correlation Studies

Kolonel and co-workers[31,43,44] studied the statistical association of selected dietary parameters with the incidence of prostate cancer in Hawaii. Preset criteria were defined for the significance of statistical associations, based on magnitude of the effect and consistency and strength of the association. According to these criteria significant positive associations were found for animal fat (r = 0.90), saturated fat (r = 0.87), total protein (r = 0.78), and animal protein (r = 0.83).

Gaskill and co-workers[45] presented correlations between prostatic cancer mortality and milk and egg demand within the U.S. For milk demand a negative zero-order correlation coefficient of -0.464 ($p < 0.001$) was found.

Blair and Fraumeni[46] suggested a positive association for white, but not for black men, between the mortality figures from seven regions in the U.S. with the consumption of fatty foods in four regions of the U.S. However, comparison of their mortality figures with the detailed graphical presentation by Gaskill et al.[45] of the same data on the consumption of fatty foods, sources of carbohydrate, and the intake of alcohol and total energy, does not reveal consistent associations with any of the commodities covered.

In a state by state comparison in the U.S. of cancer mortality and alcohol consumption, Breslow and Enstrom[19] did not find significant correlations for the use of beer, wine, and spirits with prostate cancer mortality.

Schrauzer et al.[41,42] did not find any association between state by state mortality from prostate cancer and blood selenium concentrations in pooled blood from bloodbanks in the U.S. (1959 to 1961).

In Table 3, a summary is presented of the data on intranational correlation studies.

Table 2
SUMMARY OF INTERNATIONAL CORRELATION STUDIES[a]

Foods/nutrients	Correlation coefficients	Ref.	Association
	Foods		
Meat			
All meats	0.60, 0.74, 0.56	85, 88, 89	Positive
Cattle meat	0.75, 0.47	88, 87	Positive
Pork meat	0.41, 0.48	87, 88	Positive trend
Poultry	0.20	88	— (L)
Edible fats and oils	0.70, 0.73	85, 88	POSITIVE
Milk	0.66, 0.70, 0.35, 0.52	85, 88, 89, 87	Positive
Eggs	n.s., 0.50, 0.30, 0.48	85, 88, 89, 87	Positive trend
Fish	n.s., −0.09, −0.31	85, 88, 89	—
Cereals	−0.60, −0.68, −0.77	85, 88, 89	Negative
Bread cereals	0.28	89	— (L)
Wheat	0.32	88	— (L)
Rice	−0.72	88	Negative (L)
Maize	−0.53	87	Negative (L)
Potatoes	0.54	88	Positive (L)
Sugar(s)	0.63, 0.67, 0.62	85, 88, 89	Positive
Vegetables	0.09, −0.40	88 89	—
Pulses	−0.63, −0.66	88, 87	Negative
Pulses, nuts, seeds	−0.59	85	Negative (L)
Fruits	n.s., 0.38, 0.16	85, 88, 89	—
Citrus fruits	0.32	89	— (L)
	Nutrients and Energy		
Fat			
Total fat	0.74, 0.70, 0.72, 0.89	85, 89, 87, 87	POSITIVE
Animal fat	0.51	87	Positive (L)
Protein			
Total protein	0.50, 0.52	85, 87	Positive
Animal protein	0.67	85	Positive
Sugar	0.63, 0.67, 0.62	85, 88, 89	Positive
Starch	n.s., 0.36	85, 88	—
Calories	0.61	85	Positive (L)
Animal calories	0.76	87	Positive (L)
	Beverages		
Coffee	0.57[b], 0.63, $p<0.001$[c], 0.70	85, 89, 58, 90	Positive
Tea	n.s., −0.25, n.s.	85, 89, 58	—
Beer	0.42, 0.44	89, 87	Positive trend
Wine	−0.07	89	— (L)
Hard liquor	0.25	89	— (L)

Note: POSITIVE, strong positive association; positive/negative, weak to moderate positive/
negative association; —, no association; (L), limited evidence (one study only); n.s.,
no correlation coefficient given; only stated that no significant association was found.

[a] Mortality data were in all but one study[40] derived from Segi[5] over the period 1964 to 1965.
Takahashi[40] used a similar source over the period 1956 to 1959. Armstrong and Doll[35] used
incidence data from "Cancer Incidence in Five Continents", Volumes 1 and 2. Data on per
capita consumption of various food commodities were obtained from the FAO Food Balance
Sheets over the period 1963 to 1965[35] or 1964 to 1966,[36-38] or from OECD publications
covering 1960 to 1965.[39] For data on coffee and tea consumption, statistical information
from the U.N.O. over 1965 to 1966 was used by Stocks[18] and Takahashi[40] for 20 countries,
and Carroll and Khor,[36] Correa[17] and Howell[38] covered 41 countries.

[b] This correlation coefficient was 0.31 after correction for fat consumption.

[c] Only *p*-value was given, and no correlation coefficient.

Table 3
SUMMARY OF THE RESULTS OF WITHIN-COUNTRY CORRELATION STUDIES

Study	Population/dietary information	Mortality/incidence	Association		
			Positive	Negative	No association
Kolonel et al.[31,43,44]	Representative samples of the 5 major ethnic groups in Hawaii (n = 200— 922, total n = 2247); Interview on frequency of use and amounts consumed of 83 food items in 1977—1979	Incidence 1973—1977	Animal fat, saturated fat, total protein, animal protein	—	Total-, vegetable-, and unsaturated fat, meat-, fish-, and dairy fat, cholesterol, eggs, vitamins A and C
Gaskill et al.[45]	Per capita demand for eggs and milk estimated from the 1965—1966 Household Food Consumption Survey, for 48 states in the U.S.	Death rates in 48 states in the U.S.		Milk	Eggs
Blair and Fraumeni[46]	Weekly consumption of fatty foods estimated from the 1965—1966 Household Food Consumption Survey for 7 regions in the U.S.	Death rates in 7 regions in the U.S.	(Fatty foods)[a]		
Breslow and Enstrom[19]	Alcohol consumption per state in the U.S. estimated from 1960 tax figures	Death rates in the U.S. per state, 1950—1967			Beer, wine, spirits
Schrauzer et al.[41,42]	Whole blood selenium concentrations in pooled samples from blood banks in 19 states in the U.S.	Death rates in 19 states in the U.S. 1956—1961			(Selenium)[a]

a   Weak or unconvincing results are presented in parentheses.

2. Retrospective (Case-Control) Studies

Graham and co-workers[47] studied white prostatic cancer patients in Buffalo, N.Y. They calculated indices for the intake of total fat, animal fat, vitamin C, and vitamin A, including all sources of vitamin A. These estimations are probably not very accurate, since the authors stated that some potentially essential information was not sampled. In their analysis, prostatic cancer risk increased significantly with increasing consumption of meats and fish, "all fats", animal fats, vitamin A, and vitamin C, when examining patients and controls of all ages. For the age group under 70 years, there were no such patterns. They were only found for patients of 70 years and over, but were less pronounced than for all ages together, and were not found for the category "all fats". None of the significant associations between frequency or amount and relative risk show a perfect dose-response relationship. This was particularly clear for animal fats, and for the vitamin A index, where risk actually went down for the highest intake group for both age groups, after an initial rise in the lower intake groups. The authors did not report on specific food, except for cruciferous vegetables (in general, these include: various cabbages, cresses, rutabaga/turnips, kohlrabi, radish) for which no association with prostatic cancer was found.

Results from a case-control study in Japan are reported by Mishina et al.[21] Cases consumed seafood more frequently than controls (relative risk, RR = 1.97, $p < 0.05$). Cases tended to consume less green/yellow vegetables (RR = 1.97, $0.05 < p < 0.10$) and more alcohol (RR = 1.73, $0.05 < p < 0.10$).

Preliminary results from a case-control study among the five major ethnic groups in Hawaii, are reported by Kolonel and co-workers.[31] For men of 70 years of age and older, relative risks were consistently higher than 1.0 for the consumption of total fat, saturated fat, and unsaturated fat, showing positive dose-response gradients. These findings were not statistically significant. For men less than 70 years, no association with risk was found for these dietary factors, and particularly no suggestion of dose-response gradients.

Another preliminary report concerned a study by Rotkin[22] in two U.S. hospitals. Dietary information (frequencies of use) was obtained by an interview administered questionnaire on a limited number of mainly fatty foods. Cases at both institutions consumed significantly more eggs ($p < 0.001$), and margarine ($p = 0.04$) than controls and tended to consume more cheese ($p = 0.09$) and butter ($p = 0.09$). Cases also seemed to consume more of what was called "all creams" (both institutions) and, in one hospital only, more beef and pork, but these differences were not statistically significant. The authors concluded that cases tended to consume more fatty foods than did controls, in particular more eggs, but not more milk. The evidence presented, however, is not convincing.

More extensive dietary information was obtained in a preliminary reported case-control study in the U.S. by Schuman and co-workers.[24] Food items that showed a consistent negative association with prostatic cancer for both hospital and neighborhood controls were chicken, fish, cabbage, peas, carrots, and liver. There were no consistent positive associations. Our conclusions diverge somewhat from those of the authors in that we do not consider their data of sufficient consistency and strength to conclude that fatty foods in general are of potential importance for prostatic cancer risk.

In a preliminary communication, Kaul et al.[48] report on an ongoing case-control study in U.S. black men. Cases used margarine more often for cooking food ($p < 0.04$), but consumed less wine ($p < 0.01$) than controls. Calcium, potassium and iron intake was lower in cases than in controls ($p < 0.01$) as estimated from 24-hr recall data.

Preliminary data from a case-control study by Ross et al.[23] indicate that less cases consumed carrots frequently ($\geq$ 3 times per week) than controls (RR = 0.4, $p < 0.05$).

In a study by Williams and Horms,[20] based on the Third National Cancer Survey, cases appeared to have consumed more wine ($p < 0.01$) than nondrinking controls.

Wynder and co-workers[17] conducted a case-control study in the New York City area.

They distinguished drinkers from nondrinking men, and did not find a difference in proportion of drinkers between cases and controls.

A summary of the methodology and results from the case-control studies is presented in Tables 4a and 4b.

3. Prospective (Cohort) Studies

Hirayama[49] is conducting a large-scale prospective study in Japan. Preliminary results of the 10-year follow-up are available. The negative association found between the frequency of use of green/yellow vegetables and prostate cancer risk was consistent for different age groups and was not affected by standardization for age and smoking. There was a clear dose-response gradient. No statistical evaluation of the data was presented. Green/yellow vegetables included cruciferous vegetables, pumpkin, green pepper, lettuce, spinach, carrots, chives, leeks, asparagus and chicory. They account for 44% of the total intake of vitamin A and carotenoids in the average Japanese diet, and for 23% of the vitamin C intake.[49]

A preliminary report of a prospective study among white SDA in California by Philips and Snowdon[50] contains information on a 21-year follow-up period. There was a slight tendency of a higher age-adjusted mortality for men using meat 4 to 7 days/week and a lower mortality for men drinking two or more cups of coffee per day.

Kark and co-workers[52] followed a cohort to males in Evans county, Georgia. Serum retinol values were measured after 14 to 16 years of frozen storage of 85 detected cancer cases and 174 age, sex, and race-matched controls. Cancer patients (all cancers) had significantly lower serum retinol levels than controls ($p < 0.003$), as was also found for white men ($p < 0.02$), and for black men ($p < 0.07$). Consistent with this finding, the eight prostate cancer deaths also had lower serum retinol levels than did controls.

In Table 5, a summary of the data from these prospective studies is given.

4. Special Populations

Prostate cancer patterns in SDA in California and Utah LDS may provide information on the importance of certain dietary factors in prostatic cancer, since these two religious groups have a lifestyle and dietary habits that differ from that of the average American.

In summary, both SDA and LDS males may have somewhat lower prostate cancer rates than the average male population in certain regions of the U.S., but not in others, whereas LDS may even show slightly higher rates in some regions.[32-34,52-54] Thus, the LDS and SDA lifestyle does not seem to influence prostatic cancer risk profoundly. The use of alcoholic beverages and smoking are proscribed for both religous groups.[32,52,54] Coffee and tea are also proscribed for LDS, while SDA tend to avoid drinking them, and rather strict sexual mores pertain to both groups.[32,52,54] In addition, active SDA are lacto-ovo-vegetarians, while LDS are not,[32-34,52-54] indicating that this dietary difference is not very important. Based on a small number of cases, a study on cancer mortality among members of a vegetarian society also indicated that prostatic cancer risk is not different in vegetarians.[55]

5. Epidemiology: Summary

In Table 6, a summary of the results of all above-described studies is presented. We have arbitrarily evaluated the strength and consistency of the data for each food item or nutrient, taking into consideration the type of studies and their weaknesses.

From the epidemiological data available, it is not clear whether only fat, or protein, or both factors, either independent or in interaction, contribute to the risk for prostate cancer. For both fat and protein, meat, fish, edible fats, and oils (including margarines and butter), milk, cheeses, and eggs are main sources. Edible fats/oils and margarine/butter together are clearly positively associated with risk for prostate cancer. When total meat, cattle meat, and pork are taken together, there is only a rather weak positive association with prostate cancer risk. This may indicate that fat is more important than protein.

Table 4a
SUMMARY OF THE METHODOLOGY OF THE CASE-CONTROL
STUDIES SUMMARIZED IN TABLE 4b

Study	Number of cases	Control		Dietary information
		Number	Type	
Graham et al.[47]	262	259	Hospital; age-matched	12—15 min interview on frequency of use of a limited number of unspecified foods; indices of intake were calculated; odds ratio gradients and statistical evaluation are presented
Mishina et al.[21]	100	100	Residence and age matched	Questionnaire, administered by interview (ca. 10 min.) on frequency of use of six (vaguely specified) food categories; relative risks (no gradients) and statistical evaluation are presented
Kolonel et al.[31] (preliminary report)	243	321	Unspecified	Consumption histories were collected by interview on frequency of use and amounts consumed of 83 (unspecified) food items; indices of intake were calculated for fats and proteins and, less reliably, for carbohydrates and vitamins A and C; relative risk gradients and statistical evaluation are presented
Rotkin[22] (preliminary report)	111	111	Hospital; age/race-matched	Interview administered questionnaire on frequency of use of a limited number of mainly fatty foods; the mean frequency of use for cases and controls and statistical evaluation are presented
Schuman et al.[24] (preliminary report)	223	223 + 223	Hospital + neighborhood; age-matched	Questionnaire (83—90% response rate) on frequency of use of 20 specified food items; odds ratio gradients, but no statistical evaluation are presented
Kaul et al.[48] (preliminary report)	80	80	Unspecified; age-matched	Questionnaire on frequency of use of unspecified number of foods; only statistical evaluation is presented

Table 4a (continued)
SUMMARY OF THE METHODOLOGY OF THE CASE-CONTROL
STUDIES SUMMARIZED IN TABLE 4b

Study	Number of cases	Control Number	Control Type	Dietary information
Ross et al.[23] (preliminary report)	110	110	Unspecified; age-matched	Interview on frequency of use of carrots; relative risk (no gradient) and statistical evaluation are presented
Williams and Horms[20]	531	ca. 1500	Nonprostatic cancer patients from the Third National Cancer Survey interview study	Interview on amount and duration of use of beer, wine and hard liquor
Wynder et al.[17]	217	200	Hospital; age/race-matched	Interview administered questionnaire on alcohol use

Vegetables, notably cruciferous vegetables, carrots, and pulses, are in general weakly negatively associated with risk. These vegetables are rich in vitamin A/carotenoids and, to a lesser extent, vitamin C. They may also contain many known cancer-inhibiting components, such as indoles, aromatic isothiocyanates, methylated flavones, and plant sterols.[56] In one of the studies, a prospective study,[49] a clear negative association was found for this type of vegetables. This study was performed in Japan, a notable low-risk country, which may lessen the impact of the findings for high-risk populations. In one case-control study[47] a positive association was found between prostate cancer risk and the estimated intake of vitamin A and vitamin C.[47] There are, however, many methodological weaknesses in this study, and, moreover, all other studies show opposite results. In our opinion, it cannot be ascertained from the epidemiological data available today, whether vitamin A and vitamin C are related to prostatic cancer risk in man.

B. Experimental Studies

1. In Vitro Studies

Vitamin A can counteract and reverse the hyperplastic effects of carcinogens on mouse ventral prostate explants in long-term culture. This was first shown by Lasnitski[57] when studying the effects of retinol on methylcholanthrene-induced hyperplasia in mouse ventral prostate. Incubation of prostate organ cultures for 7 to 10 days with chemical carcinogens causes hyperplastic and squamous metaplastic changes in the tissue explants and greatly enhances cell proliferation.[57-60] This occurs as reaction to both indirect-acting carcinogens, i.e., 3-methylcholanthrene and benzo(a)pyrene, and direct-acting compounds, notably *N*-methyl-*N*-nitro-*N*-nitrosoguanidine.[57-60] Retinoids counteract both effects when administered together with the carcinogen, and reverse the effects when given after treatment with the carcinogen has stopped.[57-61] Retinol (all-*trans*-retinol) and β-retinoid acid, both naturally occurring retinoids, are effective, retinoic acid being more efficient than retinol.[61] Several synthetic retinoids have similar effects, though with varying efficiency.[59-61] Carotenoids have not been studied in these systems, and only mouse ventral prostate tissue was used in these studies.

Thus, vitamin A appears to be able to maintain and restore normal differentiation and

Table 4b
SUMMARY OF THE RESULTS FROM CASE-CONTROL STUDIES, EVALUATED FOR THE STRENGTH OF THE ASSOCIATIONS FOUND

Nutrients/foods	Association			
	Positive	Possibly positive	Negative	No association
Nutrients				
Total fat	31[a]	47		
Animal fat	47			
Saturated fat	31			
Unsaturated fat	31			
Cholesterol				31
Vitamin A	47			
Vitamin C	47			
Foods				
Meat				22, 21
Cattle meat				24
Pork meat(s)				24
Chicken meat			24	
Meat + fish	47			
Fish/seafood			21, 24	
Eggs	22	24		
Milk				21, 24
Cheeses	22			
Margarine/butter	22, 48			
Ice cream		24		
Liver			24	
Green/yellow vegetables			21	
Cruciferous vegetables				47
Cabbage			24	
Carrots			23, 24	
Beans				24
Peas			24	
Tomatoes				24
Rice + cereals				21
Beverages				
Coffee				21
Tea				21
Total alcohol		21		17, 20
Wine	20		48	
Beer				20
Hard liquor				20

[a]   Reference.

adjust cell proliferation to normal rates in mouse prostate organ cultures exposed to various chemical carcinogens.[59]

2. Animal Studies
a. Animal Models for Prostatic Cancer
Events that occur in the late stages of the promotion phase are likely to be of crucial importance in prostatic carcinogenesis in man, i.e., during the progression from a nonin-

Table 5
SUMMARY OF THE METHODS AND RESULTS OF PROSPECTIVE STUDIES

Study	Number of participants	Number of prostatic cancer deaths	Dietary information	Association[a] Positive	Negative	No association
Hirayama[49]	122,261	63	Questionnaire on the frequency of use of a limited number of foods at the start of the study	—	Green/yellow vegetables	Meat, fish, milk, rice, pickles, tea, alcohol, soybean soup
Philips and Snowdon[50]	21,295	93—96	Questionnaire on the frequency of use of meat and coffee at the start of the study	(Meat)[a]	(Coffee)[a]	
Kark et al.[51]	1,484	8	Serum retinol in blood samples taken at the start of the study		(Blood retinol levels)[a]	

[a]   Doubtful associations are given in parentheses.

Table 6
SUMMARY OF THE EPIDEMIOLOGICAL EVIDENCE
FOR ASSOCIATIONS BETWEEN DIETARY FACTORS
AND PROSTATIC CANCER RISK

Association	Nutrients	Foods

Positive	Fat: total and animal fat
	Protein: total and animal
	  protein
		Meat(s) (except chicken);
		  edible fats and oils

Possibly positive	Oligosaccharides	Sugar Eggs

Negative		Vitamin A/β-carotene rich
		  foods (i.e., various vege-
		  tables and liver)

Possibly negative	Dietary fiber	Dietary fiber-rich foods Fish/seafood
Unclear	Vitamin A Vitamin C Selenium	Fruit Milk

vasive, small microcarcinoma to an invasively growing larger microcarcinoma (see Section II.C). Environmental factors, including dietary, are very likely to be operant at this stage. Therefore, it is essential to study the influence of dietary factors in an animal model that comprises these stages. Hence, the few transplantable prostatic tumor systems developed and characterized thus-far[62-64] are not relevant in this respect. Spontaneously occurring prostatic carcinomas in rats[65,66] have the disadvantage of very long latency periods which practically precludes nutritional experiments. Induction of prostate cancer in rats by hormonal manipulation[64,67] implies a profound alteration of the hormonal milieu. Therefore, they are not suitable for studying the influence of diet on prostatic carcinogenesis, since dietary influences may be mediated by the endocrine system.

In rats, Pour[68] induced squamous cell carcinomas, which are rarely seen in men. Katayama et al.[69] reported the induction of microscopic-size adenocarcinomas in rats by 3,2'-dimethyl-4-aminobiphenyl. In both cases the tumors occurred in the ventral prostate.

In our laboratory, adenocarcinomas were induced in the dorsolateral region of the prostate by a single i.v. injection of *N*-nitroso-*N*-methylurea after pretreatment with the antiandrogen cyproterone acetate followed by testosterone.[70] This pretreatment probably induces enhancement and synchronization in cell proliferation.[71] This seems essential, since pretreatment with cyproterone acetate alone did not result in tumors.[72]

Other carcinogens, notably 7,12-dimethylbenz(*a*)antracene or 3,2'-dimethyl-4-aminobiphenyl, also caused prostate cancer in this model.[72] To date this is the only induction model for prostate cancer available that seems relevant for the human disease, since metastasizing adenocarcinomas are induced in the dorsolateral region of the prostate which is homologous to the parts of the human prostate where most tumors, which are adenocarcinomas, originate from.[13,73] Further characteristics of the tumors, such as androgen dependence and histogenesis, have not yet been reported. The model seems suitable for studying dietary influences on prostatic carcinogenesis beyond the initiation phase.

2. Dietary Studies in Animal Systems

No studies have been reported using induction or spontaneous models. There is one preliminary report on the effects of dietary fat on the growth of the transplantable, androgen-dependent Dunning R3327H prostate adenocarcinoma in rats.[74] Feeding semipurified diets containing 0.5 or 20% corn oil did not result in differences in tumor-take, final weight of the tumors, or in DNA, protein, or lipid content of the tumors. The absence of an effect of dietary fat on the growth of the R3327H tumor is confirmed by preliminary data from our laboratory.[75]

C. Cadmium and Zinc

Occupational exposure to cadmium has repeatedly suggested to be associated with prostatic cancer risk, but only heavy exposure is possibly important in this respect.[11] The main source of nonoccupational exposure to cadmium for nonsmokers is through contaminated food, resulting in an average daily intake of 20 μg in the U.S. and Western Europe.[11] Schrauzer and co-workers[41] reported a positive international correlation between prostatic cancer mortality and exposure to cadmium through food products, as estimated from cadmium concentration and food consumption data from the FAO Food Balance Sheets over 1964 to 1966. In Japan, food levels were half the levels in the U.S. These findings are at variance with other data indicating that the food levels in Japan are more than twice those in the U.S.[11] Schrautzer et al.[41] did not find an association within the U.S. between prostate cancer mortality and cadmium levels in pooled blood from bloodbanks in 16 states.

Cadmium, when fed mixed through the diet up to a level of 50 ppm to rats[76] or administered via the drinking water at a level of 5 ppm to rats and mice,[77,78] is not carcinogenic. Administration by gavage, 0.35 mg cadmium once every week for 2 years to rats and for 18 months to mice, was also not carcinogenic.[79,80] Inhalation of cadmium containing aerosols causes lung tumors in rats.[81] When injected s.c. to rats, Leydig cell tumors of the testis and sarcomas at the injection site are induced.[82-84] None of these studies reported prostate tumors. Detailed microscopic examination of the prostates after s.c. injection of cadmium to rats[84] or lifetime feeding of cadmium to rats and mice[79,80] did not reveal any treatment-related lesions, and in particular no proliferative lesions.

Zinc is generally believed to be involved in prostatic carcinogenesis. The apparent daily intake of zinc is weakly positively associated with prostate cancer mortality on an international basis.[41] Concentrations of zinc in pooled blood samples from blood banks did not correlate with prostate cancer mortality in different states of the U.S.[41] Further epidemiological data are not available. There are, however, indirect indications that zinc may be related to prostate cancer. More zinc is accumulated by the human prostate than by any other tissue.[85] The dorsolateral region of the prostate in the rat contains very high amounts of zinc,[86] as well as zinc-binding proteins.[87] Zinc is essential for normal secretory function of the prostate.[88] Zinc levels in carcinomatous human prostate tissue are lower than in normal tissue.[89,90] Furthermore, zinc may affect androgen metabolism in the human prostate,[90] and in rats, zinc modulates plasma levels of hormones that are thought to be related to prostatic carcinogenesis.[91] In addition, these hormones regulate zinc-uptake in the rat dorsolateral prostate.[92] There is some evidence of zinc deficiency in a population at high risk, notably low-income elderly black men in urban areas in the U.S.[93]

It is conceivable that possible relationships of cadmium and micronutrients, such as vitamin A, selenium, and zinc, with prostatic carcinogenesis depend on interaction between these components. Cadmium exposure in rats for example, leads to alterations in vitamin A and zinc status.[94] Selenium may alleviate cadmium toxicity, and vice versa.[95] Zinc appears to play an important role in vitamin A homeostasis.[96] For mammary carcinogenesis in rats, a synergistic interaction has been observed between the protective effects of selenium and vitamin A.[97] Zinc may prevent the induction of sarcomas and Leydig cell tumors by cadmium.[83] Thus cadmium and zinc may influence prostatic carcinogenesis by interacting with nutritional factors. Other mechanisms are also conceivable, such as effects of cadmium or zinc on the endocrine system.[90,91,98]

It is to date not possible to assess whether dietary cadmium and zinc can modify prostatic carcinogenesis in men.

IV. MECHANISTIC CONSIDERATIONS

Knowledge about the mechanisms by which dietary factors may interact with prostatic carcinogenesis is scant. In theory, diet can affect a great number of processes and functions in the human body. Fat for example may influence carcinogen metabolism, the immune system, prostaglandin synthesis, endocrine systems, and membrane structure and function (e.g., membrane fluidity and permeability, membrane bound enzyme systems, receptors and antigenic determinants, sensitivity for lipidperoxidation). There is no information specifically regarding the possible role of fat or protein in prostatic carcinogenesis.

A. Vitamin A

Vitamin A inhibits the transforming effects of chemical carcinogens in prostate explant cultures. Culturing prostate explants in vitamin A-free medium induces similar morphological changes as the presence of carcinogens.[99] Thus, vitamin A maintains normal differentiation of prostate epithelium and confines the rate of cell proliferation within normal boundaries.[59] Vitamin A is believed to be able to modify gene expression by binding to specific intracellular receptor proteins and subsequent interaction of the retinol-receptor complexes with the DNA, either directly or via interactions with proteinkinases.[100] Interestingly, the presence of a retinol-binding protein has been reported in normal dorsolateral, but not in ventral prostate lobes of the rat, as well as in several sublines of the Dunning transplantable prostate carcinoma.[101]

B. The Endocrine System, Nutrition, and Prostate Cancer

The prostate is a hormone-dependent gland, and neoplastic prostate tissue also is often temporary androgen dependent.[102] Androgenic and estrogenic steroids, prolactin, growth hormone, and possibly also insulin and corticosteroids all play a more or less important role in normal development and function of the prostate gland.[102] These hormones are also believed to play a role in prostatic carcinogenesis.[102] Studies on the endocrine status of prostate cancer patients in comparison with that of controls have thus far not yielded a clear insight in the etiological importance of hormones.[102] Also no clear picture arises from studies on the endocrine status in populations at different risk for prostate cancer.[12,102-107]

In studies in men by Hill and co-workers, the hypothesis was explored that diet affects hormonal systems that may be related to the development of cancer of the prostate.[103-107] Their data indicated that changing from a Western diet to a vegetarian diet, and vice versa is related to changes in testicular activity and in metabolism and clearance of androgens and estrogens.[103-107] Furthermore, old men (> 55 years) seem to respond differently from younger men, which may be related to changes in the feedback control of plasma hormone levels occurring around 60 years of age.[105] Also, similar diet changes may affect plasma prolactin levels.[106]

Prolactin release in men is stimulated by consuming a meal.[108] This postprandial response lasts a few hours and is especially triggered by high-protein content meals and not by high-fat or high-carbohydrate (glucose) content meals.[108]

Daily vitamin E supplementation for 2 to 8 weeks in men has been shown to increase basal testosterone plasma concentrations, both free and total, and to enhance the testosterone response to human chorionic gonadotrophin.[109]

A number of mainly preliminary reports are available of studies on dietary modification of plasma levels of hormones in rats.[109-117] Cohen found elevated plasma levels of prolactin in rats maintained on a high-fat content diet (lard).[110] Preliminary results from our laboratory[111,112] and from Clinton et al.[116] and Mulloy et al.[117] did not confirm this finding. Also, prolactin clearance was not affected by dietary fat.[116] Perhaps differences in the method of blood sampling may account for this discrepancy, i.e., using ether anesthesia or decapitation after habituation to handling, as was suggested for female rats.[118]

Furthermore, dietary protein and carbohydrate, and caloric restriction may influence plasma hormone levels in rats.[111,113-115] We found that in general differences in dietary composition that were associated with differences in plasma levels of sex steroids and prolactin were more pronounced during puberty and adolescence than at later age.[111-114] The levels of estradiol-17β and in particular dihydrotestosterone were often affected by dietary differences, but those of testosterone and other hormones only occasionally.[111-114] This may indicate that dietary factors often influence androgen- and estrogen-metabolizing enzymes.

Vitamin E deficiency lowered plasma concentrations of LH, FSH, and testosterone in male rats, while vitamin E supplementation increased plasma levels of these hormones.[109] This indicates that dietary vitamin E can influence pituitary function in rats.

In conclusion, diet can affect hormonal status in men and rats and does so in a rather complex way, but from the scarce data available no detailed conclusions can be drawn.

V. CONCLUDING REMARKS AND RECOMMENDATIONS FOR FUTURE RESEARCH

A. Concluding Remarks

Knowledge about the epidemiology of cancer of the prostate is limited, and experimental data on the etiology of the disease are minimal. Epidemiological studies have usually been fairly weak in methodology, and experimental research has been impeded by lack of suitable animal models.

Familial aggregation of risk is a consistent, but not very strong finding. Familial effects on plasma levels of androgenic steroids[119] suggest a hormonal mechanism mediating this genetic effect. No endocrine differences, however, were found between black and white North American men.[103,104] The cause of the impressive black/white difference in prostatic cancer rates in the U.S. is not clear and warrants further investigation.

Sexual factors may play a role in prostatic carcinogenesis, but it is obscure which specific factors are important, and how they exert their effect. Consistent is the association of risk with a history of veneral disease. The association with marital status is, in our feeling, a reflection of the importance of sexual factors. All associations found for sexual factors, however, are rather weak.

With respect to dietary factors, fat, protein, and vitamin A are of interest. There is a remarkable consistency among different kinds of epidemiological studies in the positive associations with prostatic cancer risk found for fat and protein. Fat is more consistently and strongly associated with prostatic cancer risk and thus seems to be more important than protein. The observation that vitamin A counteracts carcinogen-induced morphological alterations in prostate organ cultures strongly supports the assumption that vitamin A is a major factor in the negative associations found between the consumption of certain vegetables and risk for prostate cancer. Nevertheless, other vegetable factors, such as β-carotene, indoles, flavones, and others, may be important, though perhaps to a lesser degree than vitamin A. In general, the associations found between these dietary factors and risk for prostate cancer are consistent and fairly strong. Because of the apparent inverse relation between fat/protein and vitamin A/β-carotene often found in correlation studies regarding prostatic cancer risk, it is conceivable that only one of these factors is causally important.

Other dietary factors may also be important in prostatic carcinogenesis. Epidemiology suggests that sugar and egg consumption are weakly positively associated with prostate cancer risk, and dietary fiber and fish inversely. There are very weak epidemiological indications that selenium is inversely related to risk for prostate cancer. Interestingly, selenium is a powerful inhibitor of mammary carcinogenesis in rats.[97] Exposure to dietary zinc and cadmium is perhaps also of significance for prostatic carcinogenesis.

Diet may affect prostatic carcinogenesis in men by influencing endocrine systems, as indicated by the studies by Hill and co-workers.[103,107] In these studies, old South African black men (60 to 73 years) showed a different endocrine response to a dietary change than did young men (40 to 55 years). In two case-control studies older (70 years and over) and younger patients (< 70 years) were examined separately.[31,47] Interestingly, in these studies the associations of dietary variables with risk were only found for the older group and not for younger cases. Zumoff and co-workers[120] observed that older (65 years and over) prostatic cancer patients differed in their endogenous hormonal patterns from the younger patients. They proposed a "two-disease" theory of prostate cancer as has been suggested for breast cancer.[121] The results of the above-mentioned studies strongly support this view. Furthermore, they indicate that dietary factors may be selectively important for the development of prostate cancer in old men.

Dietary factors seem not as strongly related to prostate cancer as they are to breast cancer or colon cancer, as argued in Section III. Other factors are likely to be also relatively important.

Dietary factors are in our opinion more important than sexual factors, which are in turn probably of more importance than genetic factors. Cell proliferative activity in prostatic epithelium is possibly of more significance in inducing prostatic cancer then chemical or transmissible factors in the environment. Arguments for this view are derived from our observations in rats[70,72] and the consistent association of venereal disease with prostate cancer risk (see Section II D).

We propose that prostatic carcinogenesis in men is a distinctly multifactorial process, in which the following influences are of major importance: (1) genetic factors that influence susceptibility for the disease, perhaps through endocrine mechanisms; (2) sexual factors and possibly dietary factors that influence sensitivity to initiating events, primarily by affecting cell turn-over (hormones, venereal disease, prostatitis) and secretory activity of the prostate gland; (3) dietary factors, and to a lesser extent sexual factors, that influence the promotion phase and in particular progression from low grade malignant prostatic microcarcinomas to invasively growing prostatic cancers, and that are primarily important in men over 60 years of age; (4) interactions between dietary, sexual, genetic factors and possibly chemical and transmissible factors, that may predispose U.S. black men and explain increasing time trends.

A summary of the above etiological hypothesis is given in Figure 1.

B. Recommendations for Future Research

Case-control and prospective studies especially geared to investigate dietary factors are needed. For these studies, improvement and validation of the methodology of assessing food consumption are of crucial importance. In particular, when indices of nutrient intakes are to be calculated, the selection of the food items covered should not be limited. The use of prospective studies in high-risk populations, e.g., U.S. black men, may improve their efficiency. In addition, it could be considered in epidemiological studies to use the prevalence of latent carcinoma of the prostate as "endpoint" rather than prostatic cancer incidence or mortality. Investigating populations at different risk but with the same genetic background, or populations that differ genetically but not in risk could provide very useful information on the interaction of genetics and lifestyle. In all studies sexual, occupational, genetic, and other possibly confounding variables should be covered.

Dietary effects on relevant endocrine parameters should be studied in more depth, with emphasis on standardization of the experimental procedures. In order to understand dietary influences on the prostate gland, dietary effects on the composition (e.g., hormones, zinc) of the prostatic fluid may be studied.

In future research, not only fat, protein, and vitamin A should receive attention, but also other macro- and micronutrients, as well as their interactions should be studied.

Further investigations using prostate explant cultures are needed. Modifying effects of

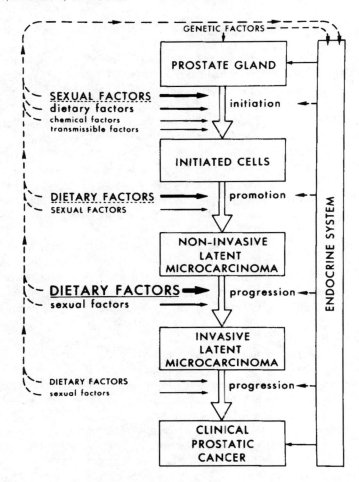

FIGURE 1.   Hypothesis of the etiology of prostatic cancer. (Size of arrows and characters indicate the importance of each factor, while interrupted arrows indicate more hypothetical associations than solid arrows.)

various micronutrients in these systems can be studied by adding them to the medium, but perhaps also by using prostates from animals that were kept on different diets. Finally, the further development of relevant animal models, in particular induction models, is warranted, including proper characterization of the models.

ACKNOWLEDGMENTS

The research presented from the laboratory of the author was supported by the Netherlands Cancer Foundation. The author wishes to thank Drs. G. S. J. Bunnik, L. A. Cohen, R. B. Hayes, J. A. Joles, and R. Kroes for their comments and Ms. J. de Bruin and N. Polter for their typing assistance.

REFERENCES

1. **White, J. E., Entecline, J. P., Alam, Z., and Moore, R. M.,** Cancer among blacks in the U.S.A. — recognizing the problem, in *Cancer Among Black Populations,* Mettlin, C. and Murphy, G. P., Eds., Alan R. Liss, New York, 1980, 35.
2. **Schuman, L. M. and Mandel, J. S.,** Epidemiology of prostatic cancer in blacks, *Prev. Med.,* 9, 630, 1980.
3. **Mandel, J. S. and Schuman, L. M.,** Epidemiology of cancer of the prostate, in *Reviews in Cancer Epidemiology,* Vol. 1, Lilienfeld, E. M., Ed., Elsevier, Amsterdam, 1980, 1.
4. **Segi, M.,** *Age adjusted deaths rats for cancer for selected sites (A-classification) in 52 countries in 1973,* Segi Institute Cancer Epidemiology, Nagoya, 1978.
5. **Segi, M.,** *Age adjusted death rates for cancer for selected sites (A-classification) in 40 countries in 1976,* Segi Institute Cancer Epidemiology, Nagoya, 1981.
6. **Waterhouse, J., Muir, C., Correa, P., and Powell, J.,** Cancer Incidence in Five Continents, Vol. 3, International Agency Research Cancer, Lyon, 1976.
7. **Waterhouse, J., Muir, C., Correa, P., and Powell, J.,** Cancer Incidence in Five Continents, Vol. 4, International Agency Research Cancer, Lyon, 1982.
8. **Murphy, G. P., Ed.,** *Prostatic Cancer,* PSG Publishers, Littleton, 1976.
9. **Greenwald, P.,** Prostate, in *Cancer Epidemiology and Prevention,* Schottenfeld, D. and Fraumeni, J. F., Eds., W. B. Saunders, Philadelphia, 1982, 839.
10. **Winkelstein, W. and Ernster, V. L.,** Epidemiology and etiology, in *Prostatic Cancer,* Murphy, G. P., Ed., PSG Publishers, Littleton, 1979, 1.
11. **Piscator, M.,** Role of cadmium in carcinogenesis with special reference to cancer of the prostate, *Environ. Health Perspect.,* 40, 107, 1981.
12. **Jackson, M. A., Ahluwalia, B. A., Herson, J., Heshmat, M. Y., Jackson, A. G., Jones, G. W., Kapoor, S. K., Kennedy, J., Kovi, J., Lucas, A. O., Nkposong, E. O., Olisa, E., and Williams, A. O.,** Characterization of prostatic carcinoma among blacks: a continuation report, *Cancer Treat. Rep.,* 61, 167, 1977.
13. **Breslow, N., Chan, C. W., Dhom, G., Drury, A. B., Franks, L. M., Gellei, B., Lee, Y. S., Lundberg, S., Sparke, B., Sternby, N. H., and Tulinius, H.,** Latent carcinoma of prostate at autopsy in seven areas, *Int. J. Cancer,* 20, 680, 1977.
14. **Yatani, R., Chigusa, I., Akazaki, K., Stemmermann, G. N., Welsh, R. A., and Correa, P.,** Geographic pathology of latent prostatic carcinoma, *Int. J. Cancer,* 29, 611, 1982.
15. **Akazaki, K. and Stemmermann, G. M.,** Comparative study of latent carcinoma of the prostate among Japanese in Japan and Hawaii, *J. Natl. Cancer Inst.,* 50, 1137, 1973.
16. **Guileyardo, J. M., Johnson, W. D., Welsh, R. A., Akazaki, K., and Correa, P.,** Prevalence of latent prostate carcinoma in two U.S. populations, *J. Natl. Cancer Inst.,* 65, 311, 1980.
17. **Wynder, E. L., Mabuchi, K., and Whitmore, W. F.,** Epidemiology of cancer of the prostate, *Cancer,* 28, 344, 1971.
18. **Stocks, P.,** Cancer mortality in relation to national consumption of cigarettes, solid fuel, tea and coffee, *Br. J. Cancer,* 24, 215, 1970.
19. **Breslow, N. E. and Enstrom, J. E.,** Geographic correlations between cancer mortality rates and alcohol-tobacco consumption in the U.S.A., *J. Natl. Cancer Inst.,* 53, 631, 1974.
20. **Williams, R. R. and Horm, J. W.,** Association of cancer sites with tobacco and alcohol consumption and socioeconomic status of patients; interview study from the third national cancer survey, *J. Natl. Cancer Inst.,* 58, 525, 1977.
21. **Mishina, T., Watanabe, H., Araki, H., Miyakoda, K., Fujiwara, T., Kobayashi, T., and Maegawa, M.,** High risk group for prostatic cancer by matched pair analysis, *Nippon Hinyokika Gakkai Zasshi,* 72, 1256, 1981.
22. **Rotkin, E. K.,** Studies in the epidemiology of prostatic cancer: expanded sampling, *Cancer Treat. Rep.,* 61, 173, 1979.
23. **Ross, R. K., Paganini-Hill, A., and Henderson, B. E.,** The etiology of prostate cancer: what does the epidemiology suggest?, *Prostate,* 4, 333, 1983.
24. **Schuman, L. M., Mandel, J. S., Radke, A., Seal, U., and Halberg, F.,** Some selected features of the epidemiology of prostatic cancer: Minneapolis-St. Paul, Minnesota case-control study, 1976—1979, in *Trends in Cancer Incidence,* Magnus, K., Ed., Hemisphere, Washington, D.C., 1982, 345.
25. **Schuman, L. M., Mandel, J., Blackard, C., Bauer, H., Scarlet, J., and McHugh, R.,** Epidemiologic study of prostatic cancer: preliminary report, *Cancer Treat. Rep.,* 61, 181, 1977.
26. **Rotkin, I. D., Moses, V. K., Kaushal, D., Cooper, J. F., Osburn, W. C., and Benjamin, J. A.,** Groups of variables available for prevention of prostatic cancer, *Cancer Detect. Prev.,* 2, 517, 1979.

27. **Jackson, M. A., Aluwalia, B. S., Attah, E. B., Connally, C. A., Herson, J., Heshmat, M. Y., Jackson, A. G., Jones, G. W., Kapoor, S. K., Kennedy, J., Kovi, J., Lucas, A., Nkposong, E. O., Olisa, E., and Williams, A. O.,** Characterization of prostatic carcinoma among blacks: a preliminary report, *Cancer Chemother. Rep.,* 59, 3, 1975.

28. **Isaace, J. T.,** Prostatic structure and function in relation to the etiology of prostatic cancer, *Prostate,* 4, 351, 1983.

29. **Berg, J. W.,** Can nutrition explain the pattern of international epidemiology of hormone-dependent cancers?, *Cancer Res.,* 35, 3345, 1975.

30. **Wynder, E. L. and Hirayama, T.,** Comparative epidemiology of cancers of the United States and Japan, *Prev. Med.,* 6, 567, 1977.

31. **Kolonel, L. N., Hankin, J. H., Nomura, A. M. Y., and Chu, S. Y.,** Dietary fat intake and cancer incidence among five ethnic groups in Hawaii, *Cancer Res.,* 41, 3727, 1981.

32. **Phillips, R. L.,** Role of life-style and dietary habits in risk of cancer among Seventh Day Adventists, *Cancer Res.,* 35, 3513, 1975.

33. **Phillips, R. L., Kuzma, J. W., and Lotz, T. M.,** Cancer mortality among comparable members versus non-members of the Seventh-Day Adventists Church, in *Cancer Incidence in Defined Populations,* Cairns, J., Lyon, J. L., and Skolnick, M., Eds., Cold Spring Harbor Lab, Cold Spring Harbor, N.Y., 1980, 93.

34. **Lyon, J. L., Gardner, J. W., and West, D. W.,** Cancer risk and life style: cancer among Mormons from 1967—1975, in *Cancer Incidence in Defined Populations,* Cairns, J., Lyon, J. L., and Skolnick, M., Eds., Cold Spring Harbor Lab, Cold Spring Harbor, N.Y., 1980, 3.

35. **Armstrong, B. and Doll, R.,** Environmental factors and cancer incidence and mortality in different countries with special reference to dietary practices, *Int. J. Cancer,* 15, 617, 1975.

36. **Carroll, K. K. and Khor, H. T.,** Dietary fat and cancer, *Prog. Biochem. Pharmacol.,* 10, 308, 1975.

37. **Correa, P.,** Epidemiological correlations between diet and cancer frequency, *Cancer Res.,* 41, 3685, 1981.

38. **Howell, M. A.,** Factor analysis of international cancer mortality data and *per capita* food consumption, *Br. J. Cancer,* 29, 328, 1974.

39. **Schrauzer, G. N.,** Cancer mortality correlation studies. II. Regional associations of mortalities with the consumptions of foods and other commodities, *Med. Hypoth.,* 2, 39, 1976.

40. **Takahashi, E.,** Coffee consumption and mortality for prostate cancer, *Tohoku J. Exp. Med.,* 82, 218, 1964.

41. **Schrauzer, G. N., White, D. A., and Schneider, C. J.,** Cancer mortality correlation studies. IV. Associations with dietary intakes and blood levels of certain trace elements, notably Se-antagonists, *Bioinorg. Chem.,* 7, 35, 1977.

42. **Schrauzer, G. N., White, D. A., and Schneider, C. J.,** Cancer mortality correlation studies. III. Statistical associations with dietary selenium intakes, *Bioinorg. Chem.,* 7, 23, 1977.

43. **Kolonel, L. N., Hankin, J. H., Lee, J., Chu, S. Y., Nomura, A. M. Y., and Hinds, M. W.,** Nutrient intakes in relation to cancer incidence in Hawaii, *Br. J. Cancer,* 44, 332, 1981.

44. **Kolonel, L. N., Nomura, A. M. Y., Hinds, M. W., Hirohata, T., Hankin, J. H., and Lee, J.,** Role of diet in cancer incidence in Hawaii, *Cancer Res.,* 43, 2397s, 1983.

45. **Gaskill, S. P., McGuire, W. L., Osborne, C. K., and Stern, M. P.,** Breast cancer mortality and diet in the U.S.A., *Cancer Res.,* 39, 3628, 1979.

46. **Blair, A. and Fraumeni, J. F.,** Geographic patterns of prostatic cancer in the United States, *J. Natl. Cancer Inst.,* 61, 1379, 1978.

47. **Graham, S., Harighey, B., Marshall, J., Priore, R., Byers, T., Rzepka, T., Mettlin, C., and Pontes, J. E.,** Diet in the epidemiology of carcinoma of the prostate gland, *J. Natl. Cancer Inst.,* 70, 687, 1983.

48. **Kaul, L., Rao, M. S., Heshmat, M. Y., and Kovi, J.,** Relationship of nutrition to prostate cancer in blacks, in Proc. 12th Int. Conf. Nutr., San Diego, 1981, 164.

49. **Hirayama, T.,** Epidemiology of prostate cancer with special reference to the role of diet, *Natl. Cancer Inst. Monogr.,* 53, 149, 1979.

50. **Phillips, R. L. and Snowdon, D. A.,** Association of meat and coffee use with cancers of the large bowel, breast, and prostate among Seventh-Day Adventists: preliminary results, *Cancer Res.,* 43, 2403s, 1983.

51. **Kark, J. D., Smith, A. H., Switzee, B. R., and Hames, C. G.,** Serum vitamin A (retinol) and cancer incidence in Evans county, Georgia, *J. Natl. Cancer Inst.,* 66, 7, 1981.

52. **Phillips, R. L., Snowdon, D. A., and Brin, B. N.,** Cancer in vegetarians, in *Environmental Aspects of Cancer: the Role of Macro and Micro Components of Foods,* Wynder, E. L., Leveille, G. A., Weisburger, J. H., and Livingston, G. M., Eds., Food and Nutrition Press, Westport, Ct., 1983, 53.

53. **Jensen, O. M.,** Cancer risk among Danisch male Seventh-Day Adventists and other temperance society members, *J. Natl. Cancer Inst.,* 6, 1011, 1983.

54. **Enstrom, J. E.,** Health and dietary practices and cancer mortality among Californian Mormons, in *Cancer Incidence in Defined Populations,* Cairns, J., Lyon, J. L., and Skolnick, M., Eds., Cold Spring Harbor Lab, Cold Spring Harbor, N.Y., 1980, 69.

55. **Kinlen, L. J., Hermon, C., and Smith, P. G.,** A proportionate study of cancer mortality among members of a vegetarian society, *Br. J. Cancer,* 48, 355, 1983.
56. **Wattenberg, L. W.,** Inhibition of neoplasia by minor dietary constituents, *Cancer Res.,* 43, 1228s, 1983.
57. **Lasnitski, I.,** The influence of A-hypervitaminosis on the effect of 20-methylcholanthrene on mouse prostate glands in vitro, *Br. J. Cancer,* 9, 434, 1955.
58. **Chopra, D. P. and Wilkoff, L. J.,** Inhibition and reversal by β-retinoic acid of hyperplasia induced in cultured mouse prostate tissue by 3-methylcholanthrene or N-methyl-N-nitro-N-nitrosomethyl-guanidine, *J. Natl. Cancer Inst.,* 56, 583, 1976.
59. **Chopra, D. P. and Wilkoff, L. J.,** Activity of retinoids against benzo(a)pyrene induced hyperplasia in mouse prostate organ cultures, *Eur. J. Cancer,* 15, 1417, 1979.
60. **Lasnitski, I.,** Reversal of methylcholanthrene-induced changes in mouse prostates in vitro by retinoic acid and its analogues, *Br. J. Cancer,* 34, 239, 1976.
61. **Lasnitski, I. and Goodman, D. S.,** Inhibition of the effects of methylcholanthrene on mouse prostate in organ culture by vitamin A and its analogs, *Cancer Res.,* 34, 1564, 1974.
62. **Smolev, J. K., Heston, W. D. W., Scott, W. W., and Coffey, D. S.,** Characterization of the Dunning R3327H prostatic adenocarcinoma: an appropriate animal model for prostatic cancer, *Cancer Treat. Rep.,* 61, 273, 1977.
63. **Pollard, M., and Luckert, P. H.,** Transplantable prostate adenocarcinomas in rats, *J. Natl. Cancer Inst.,* 54, 643, 1975.
64. **Noble, R. L.,** Prostate carcinoma of the Nb rat in relation to hormones, *Int. Rev. Exp. Pathol.,* 23, 113, 1983.
65. **Ward, J. M., Reznik, G., Stinson, S. F., Lattirada, C. P., Longfellow, D. G., and Cameron, T. P.,** Histogeneis and morphology of naturally occuring prostatic carcinoma in the ACI/segHapBr rat, *Lab. Invest.,* 43, 517, 1980.
66. **Pollard, M.,** Spontaneous prostate adenocarcinomas in aged germfree Wistar rats, *J. Natl. Cancer Inst.,* 51, 1235, 1973.
67. **Pollard, M., Luckert, P. H., and Schmidt, M. A.,** Induction of prostate adenocarcinomas in Lobund Wistar rats, *Prostate,* 4, 563, 1982.
68. **Pour, P. M.,** Prostatic cancer induced in MRC rats by N-nitrosobis(2-oxopropyl)amine and N-nitrosobis(2-hydroxypropyl)-amine, *Carcinogenesis,* 4, 49, 1983.
69. **Katayama, S., Fiala, E., Reddy, B. S., Rivenson, A., Silverman, J., Williams, G. M., and Weisburger, J. H.,** Prostate adenocarcinoma in rats: induction by 3,2'-dimethyl-4-aminobiphenyl, *J. Natl. Cancer Inst.,* 68, 867, 1982.
70. **Bosland, M. C., Prinsen, M. K., and Kroes, R.,** Adenocarcinomas of the prostate induced by N-nitroso-N-methylurea in rats pretreated with cyproterone acetate and testosterone, *Cancer Lett.,* 18, 69, 1983.
71. **Tuohimaa, P.,** Control of cell proliferation in male accessory sex glands, in *Male Accessory Sex Glands,* Spring-Mills, E. and Hafez, E. S. E., Eds., Elsevier/North-Holland, Amsterdam, 1980, 131.
72. **Bosland, M. C., Prinsen, M. K., and Kroes, R.,** Chemical induction of prostatic adenocarcinomas in rats, *Proc. Am. Assoc. Cancer Res.,* 25, 103, 1984.
73. **Sandberg, A. A., Karr, J. P., and Müntzing, J.,** The prostate of dog, baboon and rat, in *Male Accessory Sex Glands,* Spring-Mills, E. and Hafez, E. S. E., Eds, Elsevier/North-Holland, Amsterdam, 1980, 566.
74. **Spriggs, C. E., Clinton, S. K., and Visek, W. J.,** The effects of dietary fat on the Dunning R3327H transplantable prostatic adenocarcinoma, *Fed. Proc.,* 42, 1313, 1983.
75. **Bosland, M. C. and Prinsen, M. K.,** unpublished observations, 1983.
76. **Löser, E.,** A 2-year oral carcinogenicity study with cadmiun in rats, *Cancer Lett.,* 9, 191, 1980.
77. **Schroeder, H. A., Balassa, J. J., and Vinton, W. H.,** Chromium, lead, cadmium, nickel and titanium in mice: effect on mortality, tumors and tissue levels, *J. Nutr.,* 83, 239, 1964.
78. **Schroeder, H. A., Balassa, J. J., and Vinton, W. H.,** Chromium. cadmium and lead in rats: effects on life-span. tumors and tissue levels, *J. Nutr.,* 86, 51, 1965.
79. **Levy, L. S. and Clack, J.,** Further studies on the effect of cadmium on the prostatic gland. I. Absence of prostatic changes in rats given oral cadmiun sulphate for two years, *Ann. Occup. Hyg.,* 17, 205,1975.
80. **Levy, L. S., Clack, J., and Roe, F. J. C.,** Further studies on the effect of cadmium on the prostate gland. II. Absence of prostatic changes in mice given oral cadmium sulphate for eighteen months, *Ann. Occup. Hyg.,* 17, 2, 1975.
81. **Takenaka, S., Oldiges, H., König, H., Hochrainer, D., and Oberdörster, G.,** Carcinogenicity of cadmium aerosols in W. rats, *J. Natl. Cancer Inst.,* 70, 367, 1983.
82. **Haddow, A., Roe, F. J. C., Dukes, C. E., and Mitchley, B. C. V.,** Cadmium neoplasia: sarcomata at the site of injection of cadmium sulphate in rats and mice. *Br. J. Cancer,* 18, 667, 1964.
83. **Gunn, S. A., Gould, T. C., and Anderson, W. A. D.,** Effect of zinc on cancerogenesis by cadmium, *Proc. Soc. Exp. Biol. Med.,* 115, 653, 1964.
84. **Levy, L. S., Roe, F. J. C., Malcolm, D., Kazantis, G., Clack, J., and Platt, H. S.,** Absence of prostatic changes in rats exposed to cadmium, *Ann. Occup. Hyg.,* 16, 111, 1973.

85. **Daniel, O., Haddad, F., Prout, G., and Whitmore, W. F.,** Some observations on the distribution of radioactive zinc in prostatic and other human tissues, *Br. J. Urol.*, 28, 271, 1956.
86. **Gunn, S. A. and Gould, T. C.,** Difference between dorsal and lateral components of dorsolateral prostate of rat in Zn^{65} uptake, *Proc. Soc. Exp. Biol. Med.*, 92, 17, 1956.
87. **Thomas, J. A., Donovan, M. P., Waalkes, M. P., and Curto, K. A.,** Interaction between protein and heavy metals in the rodent prostatic cell, in *The Prostatic Cell*, Murphy, G. P., Sandberg, A. A., and Karr, J. P., Eds., Alan R. Liss, New York, 1981, 459.
88. **Schenk, B.,** Physiologie und Pathophysiologie der Prostata und des Prostatasekrets, in *Physiologie und Pathophysiologie der Prostata*, Senge, T., Neumann, F., and Richter, K. D., Eds., Thieme, Stuttgart, 1975, 11.
89. **Habib, F. K., Hammond, G. L., Lee, I. R., Dawson, J. B., Mason, M. K., Smith, P. H., and Stitch, S. R.,** Metal-androgen interrelationships in carcinoma and hyperplasia of the human prostate, *J. Endocrinol.*, 71, 133, 1976.
90. **Habib, F. K., Mason, M. K., Smith, P. H., and Stitch, S. R.,** Cancer of the prostate: early diagnosis by zinc and hormone analysis, *Br. J. Cancer*, 39, 700, 1979.
91. **Root, A. W., Duckett, G., Sweetland, M., and Reiter, E. O.,** Effects of zinc deficiency upon pituitary function in sexually mature and immature male rats, *J. Nutr.*, 109, 958, 1979.
92. **Müntzing, J., Kirdam, R., Murphy, G. P., and Sandberg, A. A.,** Hormonal control of zinc uptake and binding in the rat dorsolateral prostate, *Invest. Urol.*, 14, 492, 1977.
93. **Wagner, P. A., Krista, M. L., Baily, L. B., Christakis, G. J., Jernigan, J. A., Aranjo, P. E., Appledorf, H., Davis, C. G., and Dinning, J. S.,** Zinc status of elderly black Americans from low-income households, *Am. J. Clin. Nutr.*, 33, 1771, 1980.
94. **Sugawara, C., Sugawara, N., and Miyake, H.,** Decrease of plasma vitamin A., albumin and zinc in cadmium treated rats, *Toxicol. Lett.*, 8, 323, 1981.
95. **Meyer, S. A., House, W. A., and Welch, R. M.,** Some metabolic relationships between toxic levels of cadmium and nontoxic levels of selenium fed to rats, *J. Nutr.*, 112, 954, 1982.
96. **Solomons, N. W. and Russell, R. M.,** The interaction of vitamin A and zinc: implications for human nutrition, *Am. J. Clin. Nutr.*, 33, 2031, 1980.
97. **Thompson, H. J., Meeker, L. D., and Becci, P. J.,** Effect of combined selenium and retinyl acetate treatment on mammary carcinogenesis, *Cancer Res.*, 41, 1413, 1981.
98. **Favino, A., Candura, F., Chiappino, G., and Cavalleri, A.,** Study on the androgen function of men exposed to cadmium, *Med. Lavoro*, 59, 105, 1968.
99. **Lasnitski, I.,** Hypovitaminosis A in the mouse prostate gland cultured in chemically defined medium, *Exp. Cell Res.*, 28, 40, 1962.
100. **Sporn, M. B. and Robberts, A. B.,** Role of retinoids in differentiation and carcinogenesis, *Cancer Res.*, 43, 3034, 1983.
101. **Gesell, M. S., Brandes, M. J., Arnold, E. A., Isaacs, J. T., Keda, H., Millau, J. C., and Brandes, D.,** Retinoic acid binding in normal and neoplastic rat prostate, *Prostate*, 3, 131, 1982.
102. **Griffiths, K., Daries, P., Harper, M. E., Peeling, W. B., and Pierrepoint, C. G.,** The etiology and endocrinology of prostatic cancer, in *Endocrinology of Cancer*, Rose, D. P., Ed., CRC Press, Boca Raton, Fla., 1979, 1.
103. **Hill, P., Wynder, E. L., Garbaczweski, L., Garnes, H., and Walker, A. R. P.,** Diet and urinary steroids in black and white North American men and black South African men, *Cancer Res.*, 39, 5101, 1979.
104. **Hill, P., Wynder, E. L., Garbaczewski, L., and Garnes, H.,** Plasma hormones and lipids in men at different risk for coronary heart disease, *Am. J. Clin. Nutr.*, 33, 1010, 1980.
105. **Hill, P., Wynder, E. L., Garnes, H., and Walker, A. R. P.,** Environmental factors, hormone status, and prostatic cancer, *Prev. Med.*, 9, 657, 1980.
106. **Hill, P. and Wynder, E. L.,** Effect of vegetarian diet and dexamethasone on plasma prolactin, testosterone and dihydroepiandrosterone in men and women, *Cancer Lett.*, 7, 273, 1979.
107. **Hill, P., Wynder, E. L., Garbaczewski, L., and Walker, A. R. P.,** Effect of diet on plasma and urinary hormones in South African black men with prostatic cancer, *Cancer Res.*, 42, 3864, 1982.
108. **Carlson, H. E., Wasser, H. L., Levin, S. R., and Wilkins, J. N.,** Prolactin stimulation by males is related to protein content, *J. Clin. Endocrinol. Metab.*, 57, 334, 1983.
109. **Umeda, F., Kato, K., Muta, K., and Ibayashi, H.,** Effect of vitamin E on function of pituitary-gonadal axis in male rats and human subjects, *Endocrinol. Jap.*, 29, 287, 1982.
110. **Cohen, L. A.,** The influence of dietary fat on plasma and pituitary prolactin in male rats, in Proc. 61th Annu. Meet. Endocrinol. Soc., Anaheim, 1979, 291.
111. **Bunnik, G. S. J., Bosland, M. C., Prinsen, M. K., and Hidajat, E.,** Dietary influences on plasma levels of testosterone, 5α-dihydrotestosterone and prolactin in male rats, *J. Steroid Biochem.*, 17, lxxxiii, 1982.

112. **Bunnik, G. S. J., Bosland, M. C., Prinsen, M. K., and van den Berg, H.,** Effects of dietary fat on plasma levels of sex steroids and prolactin in male rats of different ages, in *4th Eur. Nutr. Conf. Abstr.,* Amsterdam, 1983, 36.

113. **Bunnik, G. S. J., Bosland, M. C., Prinsen, M. K., and Floor, B.,** Effects of dietary carbohydrate and caloric restriction on plasma levels of sex steroids and prolactin in male rats of different ages, in 4th Eur. Nutr. Conf. Abstr., Amsterdam, 1983, 171.

114. **Bosland, M. C., Bunnik, G. S. J., Prinsen, M. K., and Hidajat, E.,** Effects of dietary protein on plasma levels of sex steroid and prolactin in male rats of different ages, in 4th Eur. Nutr. Conf. Abstr., Amsterdam, 1983, 171.

115. **Merry, B. J. and Holehan, A. M.,** Serum profiles of LH, FSH, testosterone, and 5α-DHT from 21 to 1000 days of age in ad libitum fed and dietary restricted rats, *Exp. Gerontol.,* 16, 431, 1981.

116. **Clinton, S. K., Mulloy, A. L., and Visek, W. J.,** Plasma prolactin (PRL) and its metabolic clearance rate (MCR) in male and female rats fed 5 and 20% dietary fat, *Fed. Proc.,* 41, 357, 1982.

117. **Mulloy, A. L., Li, P. S., Clinton, S. K., and Visek, W. J.,** Dietary protein-fat interactions on serum prolactin (PRL) and growth hormone (GH), *Fed. Proc.,* 42, 1313, 1983.

118. **Bosland, M. C. and Wilbrink B.,** The effects of dietary fat on plasma levels of prolactin and estradiol in female rats throughout the estrous cycle, in 4th Europ. Nutr. Conf. Abstr., Amsterdam, 1983, 36.

119. **Meikle, A. W. and Stanish, W. M.,** Familial prostatic cancer risk and low testosterone, *J. Clin. Endocrinol. Metab.,* 54, 301, 1982.

120. **Zumoff, B., Levin, J., Strain, G. W., Rosenfeld, R. S., O'Connor, J., Freed, S. Z., Kream, J., Whitmore, W. S., Fukushima, D. K., and Hellman, L.,** Abnormal levels of plasma hormones in man with prostate cancer: evidence toward a "two-disease" theory, *Prostate,* 3, 579, 1982.

121. **De Waard, F.,** Premenopausal and postmenopauseal breast cancer: one disease or two?, *J. Natl. Cancer Inst.,* 63, 549, 1979.

Chapter 9

DIETARY FAT AND CANCER RISK: THE RATIONALE FOR INTERVENTION

David P. Rose and Andrea P. Boyar

TABLE OF CONTENTS

I. INTRODUCTION

Interest in diet and cancer has accelerated over the past decade. This produced increased levels of funding for research projects concerned with diet, nutrition, and carcinogenesis, as well as the initiation of clinical trials examining the place of nutritional support as part of the treatment of malignant disease. Dietary recommendations and interim guidelines for reducing cancer risk in the general population have been published by the National Academy of Sciences, the National Cancer Institute, and the American Cancer Society.

Prominent among the dietary recommendations is a reduction in consumption of fat from the typical American intake of 40% of total calories to 25 to 30%. The benefits from such action, based on evidence produced by epidemiological and experimental studies and described elsewhere in this volume, include decreased risks for cancer of the breast, endometrium, colon, and perhaps prostate and ovary.

In addition to recommendations addressed to the general population, attention is also focusing on the potential value of dietary intervention in patients who are either at high risk of a specific cancer because of a preexisting disease or family history, or who have already developed a clinically manifest cancer. This chapter will discuss these issues in a critical fashion and develop some working criteria for the inclusion of patients in clinical trials of a low-fat (20% of total calories) diet. At present, the case is strongest for prescribing a low-fat diet as an adjunct to mastectomy in postmenopausal breast cancer patients. Women at high risk of breast cancer because of fibrocystic disease with epithelial atypia might also derive benefit, but the situation here is more complex because these patients are frequently premenopausal and the role of obesity and dietary fat are unclear.

Data are emerging which suggest that the fatty acid composition of dietary lipids may be as important an element in influencing cancer risk as the absolute amount of fat in the diet. It is postulated that the mechanism here involves the synthesis of prostaglandins from specific fatty acid precursors. Although there remains much to be learned in this area, the possibilities of dietary intervention based upon these concepts will be discussed in the section dealing with breast cancer.

Cancer of the colon has been associated with the consumption of high-fat, low-fiber diets, which alter the bile acid and bacterial content of the feces. Similar effects occur in patients with familial polyposis coli, and so the possibility arises of reducing colon cancer in these patients by dietary modification. At present, this is the second most likely situation in which a low-fat diet may provide clinical benefit.

Although dietary management of patients at high cancer risk may appear attractive mechanistically, its implementation requires much careful planning in order to achieve a high level of compliance. At the American Health Foundation, we have developed the basic strategies for obtaining the patients enthusiastic collaboration, as well as an extensive compilation of recipes and menus. These aspects of the dietary management are discussed in the final section of this chapter, together with approaches for verifying compliance.

II. OBESITY, DIETARY FAT, AND BREAST CANCER RISK

The epidemiological data which relate nutrition to breast cancer risk are described by Dr. Miller in Chapter 5 of this volume. From his discussion, there emerge several crucial issues which need to be addressed in planning a dietary intervention aimed at modifying risk, or the natural history of the established disease. Foremost among these is whether the consumption of a high-fat diet per se results in increased breast cancer risk rather than the presence of obesity. If this is the case, dietary intervention should be considered even though body weight may be within acceptable limits; calories derived from fat would be replaced by increasing the intake of complex carbohydrates. As to the dietary goal, the National

Research Council Committee on Diet, Nutrition, and Cancer has recommended that the American population at large reduces its fat consumption from the typical 40% of total calories to 30%.[1] However, there is a lack of any convincing evidence that this is an optimal level. Others, basing their argument on the results of international comparisons of dietary fat intake and breast cancer risk, together with studies utilizing animal models, propose a more drastic change, a reduction to 20 to 25% of calories from fat.[2,3]

In all probability, a reduction of fat to 20% of calories is an unrealistic goal for general implementation at present, and for the immediate future should be incorporated into carefully designed, controlled clinical trials involving high-risk groups with appropriate monitoring of compliance. The definition of ''high risk'' for such trials requires considerable thought. One obvious target is the remaining breast of a woman who has been treated by mastectomy for carcinoma. The development of a subsequent second breast cancer in the opposite breast occurs in some 8 to 15% of patients.[4] In addition to reducing the risk of a second primary cancer in the surviving breast, dietary modification with reduced fat consumption may decrease the incidence of local recurrence and metastatic disease after mastectomy.

III. ADVERSE INFLUENCE OF OBESITY ON PROGNOSIS AFTER MASTECTOMY

Obese patients have been found to be more likely to have a recurrence of their breast cancer after mastectomy, and to do so sooner, than nonobese women.[5-11] Recent interest in the possibility that obesity has an adverse influence on prognosis was aroused by a report by Donegan et al.,[5] who examined retrospectively the association between body weight and rate of recurrence and survival after mastectomy. The patients were stratified according to their axillary lymph node status. Body weight and prognosis were directly related for all women, but the principal determinant of outcome after surgery was whether the body weight was less or more than 130 lbs. Of particular significance was the finding that a high body weight before mastectomy exerted an adverse influence on prognosis both in terms of the disease-free interval and survival in patients without lymph node involvement. The absence of nodal involvement generally indicates a good prognosis after mastectomy; in the National Surgical Adjuvant Breast Project 87% of patients were still disease-free 5 years later,[12] and this category is excluded from what has become the standard forms of adjuvant chemotherapy. Nevertheless, as will be discussed in more detail later, a subgroup of patients without metastases to the regional lymph nodes, those with estrogen receptor negative tumors, do have a high recurrence rate.

The effect of body weight on breast cancer prognosis has been confirmed by several groups of independent investigators,[5-11] and the time is ripe to consider dietary intervention in postmastectomy patients. This entails a discussion of possible mechanisms by which obesity influences the growth of breast cancer cells and of the selection of patients likely to benefit from dietary therapy. As in the case of diet and breast cancer risk, a key issue is whether obesity per se is the factor involved, or whether this is merely an indicator of excessive fat consumption. Also, we need to know if the biological effect is mediated through endocrine changes, because if so, one criterion for patient selection may be the presence of an estrogen receptor positive tumor.[13]

A number of investigators have considered the contribution of cholesterol to the adverse effect of a high-fat diet on breast cancer risk;[14-16] the supporting evidence from clinical and experimental studies is weak at best, but Tartter et al.[7] did report that the influence of obesity plus a high serum cholesterol on disease-free survival after mastectomy was greater than either factor alone. However, although obesity on its own had a significant effect in this study, the trend observed for longer disease-free survival in patients with low serum cholesterol levels was not statistically significant.

As discussed in Chapter 5, conflicting results have been obtained from epidemiological studies designed to examine the relationship between obesity and breast cancer; more convincing in some respects is the association found with high intake of dietary fat, for it is this relationship that correlates significantly with risk of breast cancer.[17-19] Particularly noteworthy also is the rising breast cancer mortality in Japan,[20] its association with increasing consumption of animal fat,[21] and the fact that the change in risk is most apparent in postmenopausal women.[20] There are excellent correlations between breast cancer mortality rates in various areas of Japan and the consumption of total fat and pork.[20]

While much emphasis has been placed on the low incidence of breast cancer in Japan compared with the U.S., a difference which was attributed to the low levels of animal fat in the typical Japanese diet,[22,23] this international difference extends also to prognosis of the established disease. Wynder et al.[24] compared the outcome of mastectomy in Japanese and American breast cancer patients and found that the Japanese had superior 5-year survival rates, regardless of stage of disease (Figure 1). In a similar retrospective study, Sakamoto et al.[25] compared the prognosis of breast cancer patients at the Japan Cancer Institute and at Vanderbilt University Hospital in the U.S. The American patients again had a relatively poorer prognosis than did the Japanese, but further analysis showed that this applied only to the postmenopausal women (Figure 2). Limitation of the American-Japanese difference in the prognosis of surgically treated breast cancer patients to those beyond the menopause is consistent with an adverse effect of high dietary fat consumption. Additional studies are needed now that we are seeing an increasing breast cancer risk in the urban areas of Japan which is correlated with higher intakes of animal fat.

Experiments using animal models have shown that dietary fat appears to exert its effect primarily on the promotional, later stage of mammary carcinogenesis (see Chapter 6). Furthermore, a high-fat intake negates the inhibitory effect of ovariectomy on mammary tumor development after exposure of female rats to the chemical carcinogen dimethylbenz(a)anthracene,[26] a phenomenon which is in keeping with the human epidemiological data, and which perhaps at least partially models the situation in postmenopausal women with occult disease. Consistent with a promotional effect of dietary fat is the finding that in Hawaii, while *in situ* breast cancer occurs with equal frequency in Japanese-Americans and Caucasians, progression to clinically manifest disease is more common in the white population.[27] Again, this ethnic difference applies only to postmenopausal women. The likely role of dietary fat is supported by a recent dietary survey conducted in Hawaii which showed that not only was the per capita fat consumption higher in the white than in the Japanese population, but that the difference increased with advancing age.[28] Finally, within the white female population of the U.S., Brinton et al.[24] found that obesity is associated with the presence of invasive breast cancer, but not with carcinoma *in situ*.

Discussion of possible endocrine mechanisms by which obesity and dietary fat may influence breast cancer risk and prognosis has centered largely on estrogen production. After the menopause, circulating estrogens are derived very largely from the aromatization of androstenedione to yield estrone, a process which takes place predominantly in adipose tissue.[30] The conversion rate increases with age during adulthood[31] and is positively correlated with body weight.[32] There is conflict in the literature concerning plasma estrogen levels in postmenopausal breast cancer patients; some investigators reported normal levels,[33,34] while Adami et al.[35] found increased concentrations of estrone. Plasma estrone levels may also be high in patients with endometrial cancer,[36] a tumor unequivocally associated with obesity.[37,38] Both high circulating estrone concentrations and enhanced rates of androstenedione aromatization in endometrial cancer patients are due to the concurrent presence of obesity.[39] However, a difficulty arises in evoking extraglandular conversion of androstenedione to estrogen as the mechanism for obesity-associated enhancement of recurrence risk after mastectomy; there is no evidence that the conversion *rate* is increased in postmenopausal breast cancer.[34]

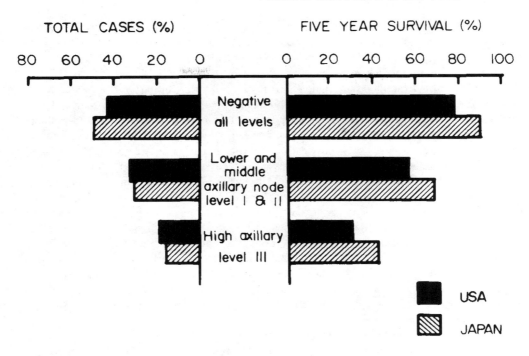

FIGURE 1. Comparison of survival rates between American and Japanese women with axillary lymph node negative or positive breast cancer. (From Wynder, E. L., Kajitani, T., Kuno, J., Lucas, J. C., DePalo, A., and Farrow, J., *Surg. Gynecol. Obstet.*, 117, 196, 1963. With permission.)

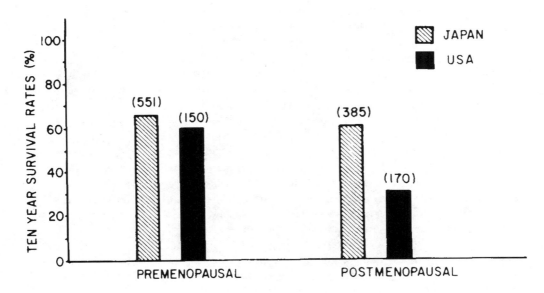

FIGURE 2. The 10-year survival rates for premenopausal and postmenopausal breast cancer patients in Japan and the U.S. (From Sakomoto, G., Sugano, H., and Hartmann, W. H., *Jpn. J. Cancer Clin.*, 25, 161, 1979. With permission.)

An alternative is that a larger pool of the precursor is available, yielding a greater amount of estrone but without an increase in the percent conversion. Adami et al.[35] did find elevated plasma androstenedione levels in their postmenopausal breast cancer patients. Also in assessing the results of a study of premenopausal and postmenopausal patients, Rose et al.[40]

performed a statistical analysis on regression lines calculated for age. The apparently higher plasma androstenedione levels in the patients compared with healthy controls did achieve borderline significance, and the lines clearly diverged with increasing age. Plasma estrone concentrations were not determined in this study.

Donegan et al.[41] examined the relationships between obesity, urinary estrogen excretion, and tumor estrogen receptor status in patients treated by mastectomy for breast cancer. There was a weak positive correlation (r = 0.17) between an obesity index (weight in pounds/height in inches) and tumor estrogen receptor content, and this was limited to the postmenopausal patients. Similarly, the urinary estrogen levels tended to increase with elevations in the obesity index, but again the relationship was a weak one as judged by the correlation coefficient (r = 0.32). The same association between obesity and estrogen receptors in postmenopausal breast cancer patients has been reported from Japan[42] and the Netherlands[43] Mason et al.[44] could not demonstrate any relationship between body weight and tumor estrogen receptor status in either premenopausal or postmenopausal patients. The reason for this discrepancy is unclear, but it should be noted that the comparison was between body weight and the proportion of tumors containing <6.0, 6.0 to 8.0, and >8.0 fmol/mg protein. These cut-off values may not have been appropriate for the purpose of the study. A fourth report, published as an abstract,[11] both confirmed the adverse influence of obesity on prognosis after mastectomy and the association between obesity and the presence of estrogen receptors in the primary tumor.

One explanation for a positive association between obesity and tumor estrogen receptor content is that the maintenance of a high level of estrogen activity after the menopause selects for hormone-dependent, receptor positive tumor cells; conversely, low estrogenic activity in postmenopausal women may only permit the proliferation of autonomous, receptor negative cells.

More work needs to be done on the effect of dietary fat intake on circulating estrogens and estrogen metabolism, as distinct from the influence of obesity. Goldin et al.[45] performed a careful investigation of ten lacto-ovo-vegetarian and ten omnivorous nonobese women, but they were premenopausal and so the results may be of limited applicability to the present discussion. In any event, the vegetarians, who obtained 30% of their total calories as fat, had a 15% reduction in plasma estrone and a 19% reduction in estradiol compared to the omnivores who derived 40% of their calories as fat. These lowered estrogen levels were not statistically significant, but there was a significant common slope ($p < 0.003$) for plasma estrogen concentration vs. fecal estrogen excretion in both groups. Thus, it appears that the higher fecal output of estrogens by the vegetarians tended to decrease their plasma levels, perhaps because of their high-fiber intake and its effect on the enterohepatic recirculation of estrogens. Alternatively, the lowered dietary fat consumption may have played a part by affecting bile acid secretions and/or altering the intestinal bacterial flora. The latter possibility is supported by the demonstration of lower fecal β-glucuronidase activity in the vegetarians, although a high-fiber intake would also have the same effect.

Further studies of dietary intakes and estrogen levels and metabolism are indicated to assist in planning clinical trials of low-fat diets in breast cancer patients and women at high risk for the disease. These would be best performed using premenopausal and postmenopausal healthy volunteers and postmastectomy breast cancer patients. The diets should be designed specifically to distinguish between the potential effects of low-fat and high-fiber intakes. As will be discussed later, the 30% of total calories as fat consumed by the lacto-ovo-vegetarians studied by Goldin et al.[45] is probably not optimal for reducing risk of recurrence after mastectomy. A more dramatic reduction to 20% of total calories is in keeping with our present concepts and is likely to have a more profound effect on the circulating estrogens.

In addition to estrogens, it is important that we obtain more information on the effects of diet on other hormones, particularly prolactin and growth hormone. Hill et al.[46] published some preliminary data from a study in which plasma prolactin levels were assayed in four women every other day through the menstrual cycle. This was done while they were consuming a typical high-fat, high-protein American diet, and again when they had been fed a diet containing no meat or meat products for 2 months. The average plasma prolactin levels were significantly lower while on the vegetarian diet in three of the women.

A number of investigators have shown that the type and not simply the quantity of dietary fat is important in influencing the development and growth of experimental mammary tumors in animal models. For example, rat mammary carcinogenesis after exposure to chemical carcinogens is stimulated by feeding a diet high in corn oil, but not by high intakes of coconut oil or tallow.[47-49] High-fat diets that promote these tumors also cause an elevation in their prostaglandin content,[45] and the stimulatory effect on carcinogenesis is suppressed by indomethacin, an inhibitor of prostaglandin synthesis.[51]

These experimental data are consistent with the hypothesis that high intakes of dietary fats such as corn oil, which are rich in linoleic acid, enhance tumorigenesis by providing precursors for prostaglandin synthesis, and that the prostaglandins thus formed stimulate tumor growth. While the mechanism by which prostaglandins might exert such an effect remain to be determined, there is evidence that they do modulate the binding of prolactin to tumor cell membrane receptors.[52]

Epidemiological studies also support the concept that dietary lipid composition as well as quantity may need to be considered in breast cancer etiology and risk modification. For example, Eskimos have a low incidence of breast and colon cancer, as well as coronary artery disease, despite their relatively high-fat intake.[53] This has been attributed to the consumption of marine oils rich in eicosapentenoic acid,[54] suggesting that it may be possible to reduce breast cancer risk in individuals at high risk by selective substitution of dietary lipids by these oils.

IV. RATIONALE FOR DIETARY INTERVENTION IN BREAST CANCER

A. Stage I and II Breast Cancer

The available data from epidemiological studies and international comparisons of postmastectomy survival indicate that modification of the total dietary fat consumption and, perhaps, of the fatty acid composition of the dietary fat, is most likely to benefit postmenopausal patients.

The question of tumor stage and estrogen receptor content is important in selecting from within the postmenopausal patient population. It seems extremely unlikely that dietary modification alone will benefit patients with a large tumor burden, regardless of whether altering fat consumption influences estrogen production, prostaglandin synthesis, or the immune response. In the U.S., it has become the standard practice to treat Stage II, axillary lymph node positive, breast cancer patients with some form of adjuvant chemotherapy. Even so, it has been reported that obesity affects adversely the prognosis of patients with Stage II disease treated by combination chemotherapy alone, or combination chemotherapy plus radiation.[55]

The presence of estrogen and progesterone-binding proteins in breast cancer tissue provides a biological marker of hormone dependence. Approximately 60% of patients with recurrent disease and estrogen receptor positive tumor tissue respond to some form of endocrine therapy; this percentage increases to 75 to 80% if both estrogen and progesterone receptors are present.[56,57] Tumor receptor status has also been correlated with prognosis after mastectomy. A number of studies showed that patients with estrogen receptor positive tumors have a better prognosis in terms of time to recurrence and disease-free survival than patients whose tumor lacks the receptor protein.[57-60] This indicator of prognosis is independent of lymph node status, but, as discussed earlier, may be predetermined by body weight.[11,41-43]

Taken together, the available evidence suggests that a low-fat diet, with caloric restriction if indicated to correct obesity, may have a place in the management of postmenopausal breast cancer patients with Stage I or II disease. An initial, controlled, clinical trial is indicated with sufficient patients to determine whether or not any therapeutic benefit is limited to patients with estrogen receptor positive tumors. Ideally, this trial would involve patients who are not receiving adjuvant chemotherapy, but this has now become accepted practice in the U.S. for patients with Stage II, axillary lymph node positive, breast cancer. A trial restricted to Stage I breast cancer would require an extremely long follow-up period given the favorable prognosis in these patients. The exception is the case of Stage I, estrogen receptor negative, patients, in whom most investigators have found the prognosis to be no better than that of patients with axillary lymph node involvement.[57-60]

Thus, the relationship between estrogen receptor status and therapeutic efficacy of a low-fat diet is the overriding issue. If only patients with estrogen receptor positive tumors benefit, the decision to be made is whether a low-fat diet is likely to be more effective, given the question of compliance, than long-term adjuvant therapy with an antiestrogen. On the other hand, if a low-fat diet proves effective in reducing risk of recurrence in patients with Stage II estrogen receptor negative breast cancer, one could feel optimistic about the outcome of a further trial involving Stage I, estrogen receptor negative patients.

While a low-fat diet may turn out to be ineffective in the case of estrogen receptor negative tumors because its mode of action involves estrogen synthesis or metabolism, this is unlikely to be true of a diet which modifies prostaglandin synthesis. The arguments in favor of initiating a clinical trial of eicosapentaenoic acid are not as well developed as those supporting a low dietary fat intervention. Further experimental and epidemiological studies should be performed to determine the relationship of dietary fatty acid constituents to premenopausal and postmenopausal breast cancer, and to the hormone dependence of tumors.

B. High-Risk Populations: Benign Breast Disease

Benign breast disease is a loose term, lacking in clinical and pathological definition, but encompassing breast cysts, fibroadenoma, intraductal papilloma, and fibrocystic disease. Yet again, fibrocystic disease of the breast is a poorly defined condition. Regardless of these problems of classification, benign breast disease is a common clinical problem and is responsible for breast biopsy in 8 to 15% of women by the age of 50 years.[61,62]

There is no doubt in the minds of most investigators that a previous diagnosis of benign breast disease is associated with an increased breast cancer risk.[63] This is not true of all forms of benign lesions, however, and further pathological considerations are necessary to ascribe a risk of later breast cancer. Black et al.[64] associated this risk with the presence of cytological atypia, a relationship confirmed on later study,[65] but not by two others in which ductal hyperplasia, with or without atypia was accompanied by the same level of increased breast cancer risk.[66,67] Page et al.[66] found that highest cancer risk to be associated with the presence of atypical lobular hyperplasia.

Several investigators have calculated the risk of subsequent breast cancer in women who previously underwent a biopsy for benign disease. Cole et al.[61] calculated the relative risk to be approximately 2.0; a similar value was also obtained by Lubin et al.[68] although here there was a nonsignificant trend of increasing relative risk with years from first biopsy, reaching a maximum of 2.9 for women with a first biopsy 11 to 15 years prior to breast cancer diagnosis. Lubin et al.[68] also examined the interactions between benign breast disease, based on previous biopsy, and other recognized risk factors. The excess breast cancer risk among women with a history of breast biopsy was enhanced by a later age at first birth, or lower parity, but, curiously, by a later menarche. The relative risk associated with benign breast disease increased from 0.9 to 1.7 to 2.2 for women aged <20, 20 to 24, and 25> years, respectively, at the time of their first completed pregnancy. Brinton et al.[69] found that breast cancer risk was increased only 1.5 times for women with a history of two breast biopsies but no family history of breast cancer. The same biopsy experience plus a history of breast cancer in a first degree relative increased the relative risk to 5.6.

A population of women at a relatively high risk for breast cancer can be identified because of a previous history of benign breast disease with a family history of breast cancer and/or nulliparity or a late first pregnancy. Whether the expression of these risk factors can be modified by altering dietary fat intake is much less certain.

There do not appear to be published data relating obesity to an increased risk of subsequent breast cancer in patients with benign breast disease, and obesity has been reported to have either no effect on the risk of benign breast disease[29] or to actually reduce the risk of symptomatic benign lesions.[62-67] In observing interactions between family history and endocrine-related characteristics, Brinton et al.[69] suggested that familial risk may be mediated by hormonal mechanisms operative early in life, effects that in all probability would also require early initiation for dietary intervention to be effective.

At the present time, there appears to be little support from epidemiological or experimental studies for a clinical trial of a low-fat diet in women at increased breast cancer risk because of a family history and/or previous benign breast disease. More basic research is needed to determine the biological mechanisms for the progression of premalignant breast disease to carcinoma *in situ* and thence to invasive cancer. Among possible hormonal factors is the influence of prolactin, growth hormone, and estrogens on cell proliferation. Studies should be performed in populations at relatively high breast cancer risk in which hormone levels are measured during carefully controlled dietary fat and protein manipulations. These should include assays of steroid and peptide hormone concentrations in both plasma and nipple aspirate breast fluids.[71,72]

V. DIETARY FAT, FIBER, AND COLON CANCER RISK

Various epidemiological studies suggest that dietary factors, specifically high dietary total fat and a deficiency of certain fibers and vegetables, are important in the etiology of colon cancer.[73-76] Also, high intake of dietary fiber and fibrous foods have been associated with a reduced risk of colon cancer in populations who consume high levels of dietary total fat.[77,78]

Studies in both animal models and in humans have demonstrated that a high-fat diet increases the concentration of fecal bile acids, and that these act as colon tumor promoters.[79,80] Also, certain bile acids enhance the formation of a mutagen which has been isolated from the feces of a population at high risk for colon cancer.[81] In contrast to these adverse effects of fat, high dietary fiber intake may reduce the carcinogenic risk by increasing stool bulk, so diluting carcinogens and promoters, binding these compounds to fibers so rendering them unavailable and accelerating their excretion in the feces.[82]

While it is extremely unlikely that patients with clinically manifest colon cancer would derive therapeutic benefit from a low-fat, high-fiber dietary regimen, clinical trials in high-risk populations would be of considerable interest. Reddy, Lipkin, and colleagues[83,84] determined fecal neutral steroid levels in symptomatic and asymptomatic patients with familial polyposis coli or hereditary colon cancer. While the total fecal neutral sterol concentrations were similar in the controls and the two groups of patients, those with familial polyposis and the affected members of colon cancer prone families excreted higher levels of coprostanol and coprostanone, bacterial metabolites of cholesterol, in their feces compared with family controls and controls other than relatives.

The effects of high-fat and low-fat, and high-fiber and low-fiber diets on fecal bile acid excretion have been studied in high- and low-risk populations for colon cancer.[85,86] Individuals at high risk and consuming a high-fat, low-fiber diet excreted higher levels of total bile acids, deoxycholic acid, and lithocholic acid than low-risk individuals consuming a low-fat diet. Also, subjects on a high-fat diet appeared to have a higher level of fecal secondary bile acids compared with those on a low-fat diet.

Related studies examined fecal bile acid excretions in populations at low colon cancer risk because of a high fiber intake despite their high-fat consumption.[87,88] It was found that the excretions of total bile acids and deoxycholic and lithocholic acids were low compared with those of a high-risk population ingesting a high-fat, but low-fiber diet.

Even this brief review of the metabolic epidemiology of colon cancer, which is discussed in detail in Chapter 4, provides convincing support of a dietary intervention trial in high-risk populations. The dietary intervention should be low in fat (20 to 25% of kcal) and high in dietary fiber (30 g). At this level of intake, mineral imbalances or deficiencies of essential fatty acids and fat-soluble vitamins from inadequate intake or impaired absorption would be unlikely.[89] The rationale for selecting this level of dietary fat and fiber is that it is within the range consumed by populations with a low incidence of colon carcinoma.[90]

VI. IMPLEMENTATION OF A LONG-TERM DIETARY INTERVENTION PROGRAM

A. Patient Entry

A pilot program is currently underway at the American Health Foundation's Mahoney Institute in which we are testing the feasibility of employing a diet low in fat (20% of calories) as adjunctive treatment for Stage I and II breast cancer. The purpose of this program is to determine whether women who were self-selected to participate could stay on the diet for at least 1 year.

At the first visit, patients are seen by their assigned nutritionist, body weight, height, and blood pressure recorded, and a blood sample obtained for the determination of serum total and high density lipoprotein cholesterol. Food-frequency questionnaires are completed, together with a 24-hr dietary recall, to help establish the patient's customary pattern of dietary intake. In addition, the patient is given verbal and written instructions on how to complete a 4-day food diary (FDFD). Detailed information is recorded in the FDFD concerning the type and quantity of foods and beverages consumed, including the fats, oils and condiments added to foods, the amount of trim on meat, skin on poultry, and the cut of meat, and the cooking methods used. Each FDFD incorporates 1 day's food intake from the weekend and 3 days during the week. Patients are asked to mail the forms back to the Institute as soon as they are completed. If the information provided on the quantities or types of food items eaten is imprecise or incomplete, the nutritionist telephones the patient to clarify the record.

A second FDFD is completed by the patient 1 week to 10 days after the initial FDFD is analyzed. From these two FDFDs the patient's average baseline intake of calories, fat, protein, carbohydrate, cholesterol, and fiber as well as the P:S ratio is calculated.

Conversion of dietary data from the FDFDs and from the 24-hr food recalls into nutrient content is performed by the use of a microcomputer-based food catalog. The machine-coded records are then transferred to a main-frame computer for analysis of nutrient content using food composition data based on USDA Handbooks No. 8, 456, and 8-1 through 8-7, and from manufacturers' information.

B. The Intervention Diet

After the baseline intake has been determined, the intervention diet is introduced to the patient and described in detail at both individual and group meetings.

The diet emphasizes the use of foods which are low in fat and rich in dietary fiber, complex carbohydrates, vitamins, and minerals. Fresh fruit and vegetables, nonfat and low-fat dairy products, whole grains and cereals, legumes, fish, skinless poultry, and lean meat form the foundation of the food plan. The distribution of calories and gram amounts of the macronutrients based on a dietary intake of 1600 kcal/day is shown in the table below:

Nutrient	Caloric %	kcal	Grams
Protein	15	240	60
Carbohydrate	65	1040	260
Fat	20	320	36
Fiber	0	0	25

In order to make the diet as easy to teach and follow as possible, the patients are asked to measure and calculate carefully in their diet only one macronutrient, fat. The amounts of fat allowed per day, the dietary fat prescription, will vary for each participant as it is based on the patient's average caloric intake as determined from the baseline data collection. The fat content of foods is measured in increments of 5 g of fat (equal to one ADA fat portion exchange), designated as a "fat portion" (FP).[91] The number of FPs that may be eaten daily by each patient is calculated according to the following formula:

$$\text{Number of FP's} = \text{Average caloric intake} \times \frac{20\% \text{ of calories}}{9 \text{ kcal per g of fat}} - \frac{5 \text{ g of fat per fat portion (FP)}}{}$$

or

Number of FP's = Kcal ÷ 225

The caloric intakes below are assigned the following number of FPs.

Caloric intake	Number of FPs
1100	5.0
1200	5.5
1400	6.0
1600	7.0
1800	8.0
2000	9.0

For example, a woman who normally eats 1800 kcal/day and who is at her ideal body weight is given a fat prescription of 8 FPs. These FPs will come from all sources of fat in the diet, both visible (butter, margarine, oil, etc.) and hidden (cheeses, meat, ice cream, etc.). The patients are instructed on the FP content of foods from the Fat Portion Exchange List which shows the amount of a particular food item that may be eaten to supply one FP. This approach enables the patient who is comfortable with controlling amounts of food and making choices based on measuring portion sizes of foods high in fat to eat almost all favorite foods, albeit in smaller quantities or not all in one day.

For patients who are unable or unwilling to count and measure fat portions carefully, and prefer instead to totally avoid foods that are high in fat, the general avoidance of all high-fat foods such as butter, margarine, mayonnaise, salad dressings, oils, and fatty meats is recommended. Patients are taught food preparation techniques that avoid the addition of fat, such as broiling, baking, poaching, and steaming. By carefully adjusting the patient's customary intake so that fat sources are gradually eliminated, the patient who needs stricter guidelines with less room for choice is accommodated.

Table 1
PLAN OF DIETARY INVERVENTION AND MEASURES OF COMPLIANCE

Dietary plan	Measures of compliance	Expected outcome
Reduce total fat intake to 20% of calories; increase dietary fiber to 25 g/day	4-day food diaries monthly Random 24-hr recalls Qualitative ratings of compliance by nutritionists Questionnaire assessing attitudes and behavioral changes Total serum cholesterol (monthly)	Reflects current dietary intake and within subject dietary changes over time 10—20% reduction from baseline measurement
Total caloric intake	Body weight (monthly)	
For women of normal weight: isocaloric intake		No change in body weight (±3 lb)
For obese women: calorie deficit no greater than 500—1000 kcal		Weight loss on average of 1—2 lb/week
For women below ideal body weight: caloric increase		Weight gain to approach ideal body weight

To ensure an adequate intake of the micronutrients, all patients are advised to supplement their diet with a vitamin and mineral preparation at nutrient levels of the Recommended Dietary Allowances (RDA). For postmenopausal women, calcium absorption may be particularly low. In order to keep bone loss at a minimum, 1000 mg of elemental calcium (125% of the RDA) is recommended, unless there is evidence of hypercalcemia (serum calcium > 11 mg/100 mℓ).

C. Nutrition Education Program

Initially, the nutrition education program is essential to provide the skills needed for planning meals and preparing food according to the dietary fat prescription. The program is composed of monthly individual counseling sessions during which the fat prescription is explained and individualized according to the patient's personal eating pattern.

Further instruction, education, and support are provided in monthly group meetings of 12 participants and 1 nutritionist. The content of group sessions includes discussion of FDFDs, accurate measurement of food portions, low-fat food preparation techniques, counting fat portions, meal planning skills, recipe demonstration, and food tasting.

At each monthly meeting, body weight is measured, blood taken for serum cholesterol, a 24-hr recall and a FDFD is collected, and a subjective evaluation of the patient's progress in following the diet is made by both the nutritionist and the patient. The nutritionist qualitatively rates the level of compliance monthly after interviewing each participant. In this way, and as outlined in Table 1, adherence to the diet is measured by laboratory as well as by nonlaboratory means.

D. Preliminary Results

Preliminary data from 13 patients who have been on the diet for at least 4 to 6 months demonstrate that fat intake drops to 20% of calories within 1 to 2 months of initial instruction. At the same time, serum cholesterol levels decrease by approximately 10%. Serum levels of prolactin and the estrogens have also decreased during the first 6 months in response to the diet.

VII. CONCLUSIONS

For women with postmenopausal breast cancer that has spread to the lymph nodes, adjuvant chemotherapy has provided only limited improvement in disease-free survival over conventional treatment. Epidemiological and animal data have shown that excess body weight and fat intake adversely influence risk for disease and recurrence after mastectomy. These data are sufficiently persuasive to justify the establishment of a multicenter trial to determine the effect of a low-fat diet on recurrence rate.

A feasibility trial performed at the American Health Foundation has already demonstrated that women with breast cancer will adhere to a diet low in total fat. It is fortunate that the National Cancer Institute has seen fit to fund a multicenter clinical trial examining the effect of a 15% fat diet on Stage II patients.

For colon cancer, evidence suggests that a low-fat, high-fiber diet would be of benefit for patients at high risk, rather than those with established disease. At the time of writing, we are not aware of any ongoing multicenter trials pursuing this provocative research question.

REFERENCES

1. Committee on Diet, Nutrition, and Cancer, National Research Council, Diet, Nutrition and Cancer, National Academy Press, Washington, D.C., 1982, 14.
2. **Wynder, E. L.**, Dietary factors related to breast cancer, *Cancer*, 46, 899, 1980.
3. **Wynder, E. L. and Cohen, L. A.**, A rationale for dietary intervention in the treatment of postmenopausal breast cancer patients, *Nutr. Cancer*, 3, 195, 1982.
4. **Slack, N. H., Bross, I. D. J., Nemoto, T., and Fisher, B.**, Experiences with bilateral primary carcinoma of the breast, *Surg. Gynecol. Obstet.*, 136, 433, 1973.
5. **Donegan, W. L., Jayich, S., Koehler, M. R., and Donegan, J. H.**, The prognostic implications of obesity for the surgical cure of breast cancer, *Breast*, 4, 14, 1978.
6. **Abe, R., Kumagai, N., Kimura, M., Hirosaki, A., and Nakamura, T.**, Biological characteristics of breast cancer in obesity, *Tohoku J. Exp. Med.*, 120, 351, 1976.
7. **Tartter, P. I., Papatestas, A. E., Ioannovich, J., Mulvihill, M. N., Lesnick, G., and Aufses, A. H.**, Cholesterol and obesity as prognostic factors in breast cancer, *Cancer (Philadelphia)*, 47, 2222, 1981.
8. **Boyd, N. F., Campbell, J. E., Germanson, T., Thomson, D. B., Sutherland, D. J., and Meakin, J. W.**, Body weight and prognosis in breast cancer, *J. Natl. Cancer Inst.*, 67, 785, 1981.
9. **Zumoff, B., Gorzynski, J. G., Katz, J. L., Weiner, H., Levin, J., Holland, J., and Kukushima, D. K.**, Nonobesity at the time of mastectomy is highly predictive of 10-year disease-free survival in women with breast cancer, *Anticancer Res.*, 2, 59, 1982.
10. **Zumoff, B. and Dasgupta, I.**, Relationship between body weight and the incidence of positive axillary nodes at mastectomy for breast cancer, *J. Surg. Oncol.*, 22, 217, 1983.
11. **Burt, J. R. F. and Schapira, D. V.**, The effect of obesity on recurrence and ER status in breast cancer patients, *Proc. Am. Soc. Clin. Oncol.*, c-6, 1983.
12. **Fisher, E. R., Redmond, C., and Fisher, B.**, Pathologic findings from the National Surgical Adjuvant Breast Project (Protocol no. 4). VI. Discriminants for five-year treatment failure, *Cancer*, 46, 908, 1980.
13. **Nordenskjold, B., Wallgren, A., Gustafsson, S., and Skoog, L.**, Steroid receptor levels and prophylactic endocrine therapy of mammary carcinoma, *Acta Obstet. Gynecol. Scand. Suppl.*, 101, 49, 1981.
14. **Lea, A. J.**, Dietary factors associated with death rates from certain neoplasms in man, *Lancet*, 2, 332, 1966.
15. **Hegsted, D. M., McGandy, R. B., Myers, S. M., and Stare, F. J.**, Quantitative effects of dietary fat on serum cholesterol in man, *Am. J. Clin. Nutr.*, 17, 281, 1965.
16. **Miller, A. B., Kelly, A., Choi, N. W., Mathews, V., Morgan, R. W., Munan, L., Burch, D., Feather, J., Howe, G. R., and Jain, M.**, A study of diet and breast cancer, *Am. J. Epidemiol.*, 107, 499, 1978.
17. **Carroll, K. K., Gammal, E. B., and Plunkett, E. R.**, Dietary fat and mammary cancer, *Can. Med. Assoc. J.*, 98, 590, 1968.
18. **Hems, G.**, Epidemiological characteristics of breast cancer in middle and late age, *Br. J. Cancer*, 24, 226, 1970.

19. **Hems, G.,** The contributions of diet and childbearing to breast cancer rates, *Br. J. Cancer,* 37, 974, 1978.
20. **Hirayama, T.,** Epidemiology of breast cancer with special reference to the role of diet, *Prev. Med.,* 7, 173, 1978.
21. **Kagawa, Y.,** Impact of westernization on the nutrition of Japanese: changes in physique, cancer, longevity and centenarians, *Prev. Med.,* 7, 205, 1978.
22. **Wynder, E. L., Bross, I. J., and Hirayama, T.,** A study of the epidemiology of cancer of the breast, *Cancer (Philadelphia),* 13, 559, 1960.
23. **Wynder, E. L. and Hirayama, T.,** Comparative epidemiology of cancers in the United States and Japan, *Prev. Med.,* 6, 567, 1977.
24. **Wynder, E. L., Kajitani, T., Kuno, J., Lucas, J. C., DePalo, A., and Farrow, J.,** A comparison of survival rates between American and Japanese patients with breast cancer, *Surg. Gynecol. Obstet.,* 117, 196, 1963.
25. **Sakamoto, G., Sugano, H., and Hartmann, W. H.,** Comparative clinicopathological study of breast cancer among Japanese and American females, *Jpn. J. Cancer Clin.,* 25, 161, 1979.
26. **Cohen, L. A., Chan, P. C., and Wynder, E. L.,** The role of a high-fat diet in enhancing the development of mammary tumors in ovariectomized rats, *Cancer (Philadelphia),* 47, 66, 1981.
27. **Ward-Hinds, M., Kolonel, L. N., Nomura, A. M. Y., and Lee, J.,** Stage specific breast cancer incidence rates by age among Japanese and Caucasian women in Hawaii (1960—1979), *Br. J. Cancer,* 45, 118, 1982.
28. **Kolonel, L. N., Hankin, J. H., Lee, J., Chu, S. Y., Nomura, A. M. Y., and Ward-Hinds, M. W.,** Nutrient intakes in relation to cancer incidence in Hawaii, *Br. J. Cancer,* 44, 332, 1981.
29. **Brinton, L. A., Hoover, R., and Fraumeni, J. F.,** Epidemiology of minimal breast cancer, *JAMA,* 249, 483, 1983.
30. **Grodin, J. M., Siiteri, P. K., and MacDonald, P. C.,** Source of estrogen production in postmenopausal women, *J. Clin. Endocrinol. Metab.,* 36, 207, 1973.
31. **Hemsell, D. L., Grodin, J. M., Brenner, P. F., Siiteri, P. K., and MacDonald, P. C.,** Plasma precursors of estrogen. II. Correlation of the extent of conversion of plasma androstenedione to estrone with age, *J. Clin. Endocrinol. Metab.,* 38, 476, 1974.
32. **MacDonald, P. C. and Siiteri, P. K.,** The relationship between the extraglandular production of estrone and the occurence of endometrial neoplasm, *Gynecol. Oncol.,* 2, 259, 1974.
33. **Wang, D. Y. and Swain, M. C.,** Hormones and breast cancer, in *Biochemistry of Women: Methods of Clinical Investigation,* Curry, A. S. and Hewitt, J. V., Eds., CRC Press, Boca Raton, Fla., 1974, 191.
34. **Thijssen, J. H. H., Poortman, J., and Schwarz, F.,** Androgens in postmenopausal breast cancer: excretion, production and interaction with estrogens, *J. Steroid Biochem.,* 6, 729, 1975.
35. **Adami, H-O., Johansson, E. D. B., Vegelius, J., and Victor, A.,** Serum concentrations of estrone, androstenedione, testosterone and sex-hormone-binding globulin in postmenopausal women with breast cancer and in age-matched controls, *Uppsala J. Med. Sci.,* 84, 259, 1979.
36. **Judd, H. L., Lucas, W. E., and Yen, S. S. C.,** Serum 17β-estradiol and estrone levels in postmenopausal women with and without endometrial cancer, *J. Clin. Endocrinol. Metab.,* 43, 272, 1976.
37. **Wynder, E. L., Escher, G. C., and Mantel, N.,** An epidemiological investigation of cancer of the endometrium, *Cancer (Philadelphia),* 19, 489, 1966.
38. **Blitzer, P. H., Blitzer, E. C., and Rimm, A. A.,** Association between teen-age obesity and cancer in 56,111 women: all cancer and endometrial carcinoma, *Prev. Med.,* 5, 20, 1976.
39. **MacDonald, P. C., Edman, C. D., Hemsell, D. L., Porter, J. C., and Siiteri, P. K.,** Effect of obesity on conversion of plasma androstenedione to estrone in postmenopausal women with and without endometrial cancer, *Am. J. Obstet. Gynecol.,* 130, 448, 1978.
40. **Rose, D. P., Stauber, P., Thiel, A., Crowley, J. J., and Milbrath, J. R.,** Plasma dehydroepiandrosterone sulfate, androstenedione and cortisol, and urinary free cortisol excretion in breast cancer, *Eur. J. Cancer,* 12, 43, 1977.
41. **Donegan, W. L., Johnstone, M. F., and Biedrzycki, L.,** Obesity, estrogen production, and tumor estrogen receptors in women with carcinoma of the breast, *Am. J. Clin. Oncol.,* 6, 19, 1983.
42. **Nomura, Y., Kanda, K., Shigematsu, T., Narita, N., Matsumoto, K., and Sugano, H.,** Relation between estrogen receptors and body weight in Japanese pre- and post-menopausal breast cancer patients, *Gann,* 72, 468, 1981.
43. **deWaard, F., Poortman, J., and Collette, B. J. A.,** Relationship of weight to the promotion of breast cancer after menopause, *Nutr. Cancer,* 2, 237, 1981.
44. **Mason, B., Holdaway, I. M., Yee, L., and Kay, R. G.,** Lack of association between weight and oestrogen receptors in women with breast cancer, *J. Surg. Oncol.,* 19, 62, 1982.
45. **Goldin, B. R., Adlercreutz, H., Gorbach, S. L., Warram, J. H., Dwyer, J. T., Swenson, L., and Woods, M. N.,** Estrogen excretion patterns and plasma levels in vegetarian and omnivorous women, *N. Engl. J. Med.,* 307, 1542, 1982.

46. **Hill, P., Chan, P., Cohen, L., Wynder, E., and Kuno, K.,** Diet and endocrine-related cancer, *Cancer (Philadelphia),* 39, 1820, 1977.
47. **King, M. M., Bailey, D. M., Gibson, D. D., Pitha, J. V., and McCay, P. B.,** Incidence and growth of mammary tumors induced by 7,12-dimethylbenz(a)anthracene as related to the dietary content of fat and antioxidant, *J. Natl. Cancer Inst.,* 63, 657, 1979.
48. **Rogers, A. E. and Wetsel, W. C.,** Mammary carcinogenesis in rats fed different amounts and types of fat, *Cancer Res.,* 41, 3735, 1981.
49. **Chan, P-C., Ferguson, K. A., and Dao, T. L.,** Effects of different dietary fats on mammary carcinogenesis, *Cancer Res.,* 43, 1079, 1983.
50. **Cohen, L. A. and Karmali, R.,** to be published.
51. **Carter, C. A., Milholland, R. J., Shea, W., and Ip, M. M.,** Effect of the prostaglandin synthetase inhibitor indomethacin on 7,12-dimethylbenz(a)anthracene-induced mammary tumorigenesis in rats fed different levels of fat, *Cancer Res.,* 43, 3359, 1983.
52. **Reichel, P., Cohen, L. A., Karmali, R. A., and Rose, D. P.,** Suppression of growth and prolactin-binding in two cultured rat mammary carcinoma cell lines by an inhibitor of prostaglandin synthesis, *Cancer Res.,* submitted.
53. **Dyerberg, J., Bang, H. O., Stoffersen, E., Moncada, S., and Vane, J. R.,** Eicosapentaenoic acid and prevention of thrombosis and atherosclerosis?, *Lancet,* 2, 117, 1978.
54. **Dyerberg, J.,** Observations on populations in Greenland and Denmark, in *Nutritional Evaluation of Fatty Acids in Fish Oil,* Barlow, S. M. and Stansby, M. E., Eds., Academic Press, New York, 1982, 245.
55. **Ahmann, D.,** Discussion, in *National Cancer Institute Consensus Conf.,* Adjuvant Chemotherapy of Breast Cancer, 1981.
56. **Edwards, D. P., Chamness, G. C., and McGuire, W. L.,** Estrogen and progesterone receptor proteins in breast cancer, *Biochim. Biophys. Acta,* 560, 457, 1979.
57. **McGuire, W. L.,** An update on estrogen and progesterone receptors in prognosis for primary and advanced breast cancer, in *Progress in Cancer Research and Therapy,* Vol. 14, Iacobelli, S., King, R. J. B., Lindner, H. R., and Lippman, M. E., Eds., Raven Press, New York, 1980, 337.
58. **Knight, W. A., Livingston, R. B., Gregory, E. J., and McGuire, W. L.,** Estrogen receptor as an independent prognostic factor for early recurrence in breast cancer, *Cancer Res.,* 37, 4669, 1977.
59. **Maynard, P. V., Blarney, R. W., Elston, C. W., Haybittle, J. L., and Griffiths, K.,** Estrogen receptor assays for primary breast cancer and early recurrence of the disease, *Cancer Res.,* 38, 4292, 1978.
60. **Croton, R., Cooke, T., Holt, S., Goerge, W. D., Nicholson, R., and Griffiths, K.,** Oestrogen receptors and survival in early breast cancer, *Br. Med. J.,* 283, 1289, 1981.
61. **Cole, P., Elwood, J. M., and Kaplan, S. D.,** Incidence rates and risk factors of benign breast neoplasms, *Am. J. Epidemiol.,* 108, 112, 1978.
62. **Hislop, T. G. and Elwood, J. M.,** Risk factors for benign breast disease: a 30-year cohort study, *Can. Med. Assoc. J.,* 124, 283, 1981.
63. **Ernster, V. L.,** The epidemiology of benign breast disease, *Epidemiol. Rev.,* 3, 184, 1981.
64. **Black, M. M., Barclay, T. H. C., Cutler, S. J., Hankey, B. F., and Asire, A. J.,** Association of atypical characteristics of benign breast leasions with subsequent risk of breast cancer, *Cancer,* 29, 338, 1972.
65. **Kodlin, D., Winger, E. E., Morgenstern, N. L., and Chen, L.,** Chronic mastopathy and breast cancer. A follow-up study, *Cancer,* 39, 2603, 1977.
66. **Page, D. L., Vander Zwaag, R., Rogers, L. W., Williams, L. T., Walker, W. E., and Hartmann, W. H.,** Relation between component parts of the fibrocystic disease complex and breast cancer, *J. Natl. Cancer Inst.,* 61, 1055, 1978.
67. **Hutchison, W. B., Thomas, D. B., Hamlin, W. B., Roth, G. J., Peterson, A. Y., and Williams, B.,** Risk of breast cancer in women with benign breast disease, *J. Natl. Cancer Inst.,* 65, 13, 1980.
68. **Lubin, J. H., Brinton, L. A., Blot, W. J., Burns, P. E., Lees, A. W., and Fraumeni, J. F., Jr.,** Interactions between benign breast disease and other risk factors for breast cancer, *J. Chron. Dis.,* 36, 525, 1983.
69. **Brinton, L. A., Hoover, R., and Fraumeni, J. F., Jr.,** Interaction of familial and hormonal risk factors for breast cancer, *J. Natl. Cancer Inst.,* 69, 817, 1982.
70. **Brinton, L. A., Vessey, M. P., Flavel, R., and Yeats, D.,** Risk factors for benign breast disease, *Am. J. Epidemiol.,* 113, 203, 1981.
71. **Wynder, E. L., Hill, P., Laakso, K., Littner, R., and Kettunen, K.,** Breast secretion in Finnish women: a metabolic epidemiologic study, *Cancer,* 47, 1444, 1981.
72. **Hill, P., Garbaczewski, L., and Wynder, E. L.,** Testosterone in breast fluid, *Lancet,* 1, 761, 1983.
73. **Wynder, E. L., Kajitani, T., Ishikawa, S., Dodo, H., and Takano, A.,** Environmental factors of cancer of the colon and rectum. II. Japanese epidemiological data, *Cancer,* 23, 1210, 1969.
74. **Haenszel, W., Berg, J. W., Segi, M., Kurihara, M., and Locke, F. B.,** Large bowel cancer in Hawaiian Japanese, *J. Natl. Cancer Inst.,* 51, 1765, 1973.

166 *Diet, Nutrition, and Cancer: A Critical Evaluation*

75. **Burkitt, D. P., Walker, A. R., and Painter, N. S.,** Effect of dietary fiber of stools and transit times and its role in the causation of disease, *Lancet,* 2, 1408, 1972.
76. **Wynder, E. L.,** The epidemiology of large bowel cancer, *Cancer Res.,* 35, 3388, 1975.
77. **Reddy, B. S., Cohen, L. A., McCoy, G. D., Hill, P., Weisburger, J. H., and Wynder, E. L.,** Nutrition and its relationship to cancer, in *Advances in Cancer Research,* Vol. 32, Weinhouse, S. and Klein, G., Eds., Academic Press, New York, 1980, 237.
78. **Jain, M., Cooke, G. M., Davis, F. G., Grace, M. G., Howe, G. R., and Miller, A. B.,** A case-control study of diet and colo-rectal cancer, *Int. J. Cancer,* 26, 757, 1980.
79. **Reddy, B. S., Watanabe, K., and Weisburger, J. H.,** Effect of high-fat diet on colon carcinogenesis in F344 rats treated with 1,2-dimethylhydrazine, methylazoxymethanol acetate and methylnitrosourea, *Cancer Res.,* 37, 4156, 1977.
80. **Reddy, B. S.,** Dietary fat and its relationship to large bowel cancer, *Cancer Res.,* 41, 3700, 1981.
81. **Van Tassell, R. L., MacDonald, D. K., and Wilkins, T. D.,** Stimulation of mutagen production in human feces by bile and bile acids, *Mutat. Res.,* 103, 233, 1982.
82. **Reddy, B. S.,** Dietary fiber and colon cancer. A critical review, in *Dietary Fiber in Health and Disease,* Vahouny, G. and Kritchevsky, D., Eds., Plenum Press, New York, 1982, 265.
83. **Reddy, B. S., Mastromarino, A., Gustafson, C., Lipkin, M., and Wynder, E. L.,** Fecal bile acids and neutral sterols in patients with familial polyposis, *Cancer,* 38, 1694, 1976.
84. **Lipkin, M., Reddy, B. S., Weisburger, J. H., and Schecter, L.,** Nondegradation of fecal cholesterol in subjects at high risk for cancer of the large intestine, *J. Clin. Invest.,* 67, 304, 1981.
85. **Reddy, B. S.,** Diet and excretion of bile acids, *Cancer Res.,* 41, 3766, 1981.
86. **Reddy, B. S.,** Dietary fat and its relationship to large bowel cancer, *Cancer Res.,* 41, 3700, 1981.
87. **Reddy, B. S., Hedges, A. R., Laakso, K., and Wynder, E. L.,** Metabolic epidemiology of large bowel cancer: fecal bulk and constituents of high-risk North American and low-risk Finnish population, *Cancer,* 42, 2832, 1978.
88. **Domellof, L., Darby, L., Hanson, D., Mathews, L., Simi, B., and Reddy, B. S.,** Fecal sterols and bacterial β-glucuronidase activity: a preliminary metabolic epidemiology study of healthy volunteers from Umea, Sweden, and metropolitan New York, *Nutr. Cancer,* 4, 120, 1982.
89. **Judd, J. T., Kelsay, J. L., and Mertz, W.,** Potential risks from low-fat diets, *Semin. Oncol.,* 10, 273, 1983.
90. **Jensen, O. M., MacLennan, R., and Wahrendorf, J.,** Diet, bowel function, and fecal characteristics and large bowel cancer in Denmark and Finland, *Nutr. Cancer,* 4, 5, 1982.
91. **Boyar, A. P. and Longbridge, J. R.,** The fat portion exchange list: a tool for teaching and evaluating low-fat diets, *J. Am. Diet. Assoc.,* 85, 589, 1985.

INDEX